W9-CCZ-709

Also by James Burke

The Knowledge Web

Connections

The Day the Universe Changed

The Axemaker's Gift
(with Robert Ornstein)

The Pinball Effect

CIRCLES

50 Round Trips through History, Technology, Science, Culture

JAMES BURKE

ILLUSTRATED BY DUŠAN PETRIČIĆ

SIMON & SCHUSTER
NEW YORK LONDON TORONTO SYDNEY SINGAPORE

SIMON & SCHUSTER
Rockefeller Center
1230 Avenue of the Americas
New York, NY 10020

First Simon & Schuster trade paperback edition 2003

Illustrations by Dušan Petričić
Designed by Karolina Harris
Manufactured in the United States of America
10 9 8 7 6 5 4 3 2 1
The Library of Congress has cataloged the hardcover edition as follows:
Burke, James, 1936–
Circles : 50 round trips through history, technology, science, culture / James Burke.
p. cm.
Includes bibliographical references.
1. Technology—History. 2. Science—History. I. Title.
T18.B86 2000
609—dc21
00-057335
ISBN 0-7432-0008-X
ISBN 0-7432-4976-3 (Pbk)

To Madeline

CONTENTS

CIRCLES

FOREWORD

I SUPPOSE THE real reason for taking an interest in history is, as some ship's navigator must have once said, you can only predict where you're going if you know where you've been. So it probably seems perverse that I have chosen to write a series of historical tales that go, as it were, round in circles, since, in some way, each of them ends where it begins. This is because of the view I take of how things happen and, specifically, of how you write about them.

First, the view. Like everything in life, the key to success, happiness, and all those other things people want lies in how good you are at prediction. The more accurately you foresee what's coming, the better you're going to be placed to (a) avoid it or (b) benefit from it. The problem is (as history shows all too painfully) that, as the great Danish physicist Niels Bohr once said, "Prediction is extremely difficult, especially about the future." This is because the future is almost never a linear extension of the present. Which all too soon becomes obvious. A few cases in point: Gutenberg thought he'd print a few Bibles, and that'd be that; in the 1940s the head of IBM said America would need about half a dozen computers, and the magazine *Popular Science* predicted they would not weigh more than 1½ tons; Alexander Graham Bell believed the telephone would be used only to tell people to expect delivery of a telegram.

Trapped inside the knowledge context of the time, people in the past were no better at second sight than we are. Which is why serendipity plays a key role in the historical sequence and the process of change. This is as true for the present as it was for the past. How many times in your own life did things come about thanks to accident of circumstance? I'll bet happenstance played a part in how you met your partner or chose your career or live where you do or a long list of other character-forming events that now make you different from anybody else. That journey from past to present, full of unexpected encounters and events along the way, has brought you to where you are and who you are at this moment, reading these words.

This is why the past is no foreign, unknown land. The people in the past were trapped in their context, just as we are in ours. Nobody back then knew what was going to come round the corner to change his or her best-laid plans.

What makes the investigation of this process so exciting (and often amusing) is that with the benefit of hindsight we can see why they thought and acted as they did. From our high ground we can also see why things almost never worked out the way they expected because, unlike us, they couldn't see round their corner.

This was why the mid-nineteenth-century coal-gas makers threw away their coal-tar by-product, unaware that within a decade it would be revealed as a valuable cornucopia of products including aspirin, antiseptic chemicals, dyes, timber preservative, and fuel.

In the eighteenth century, Italian scientist Alessandro Volta produced a gas-testing eudiometer. It consisted of a bottle into which led two wires whose ends almost touched. The experimenter filled the bottle with the gas to be investigated. A charge of electricity was then sent down one of the wires and jumped the wire gap, and the gas would explode (or not). After this kind of experiment was discontinued, the eudiometer lay around for more than a hundred years. Then it became the basic element of the spark plug.

The eighteenth-century Jacquard automated silk-weaving system, which used perforated paper as a control mechanism, inspired Herman Hollerith to develop the punch card for data processing during the 1880 U.S. census. The tabulating company he founded to exploit his idea went on to become IBM.

As I hope these essays will show, the other fascinating thing about the way events unfold is the extraordinary variety of the elements involved. In spite of the tendency in schools to segment the past into subject areas (history of chemistry, art, music, transportation, and so on) in which advances and discoveries developed the discipline into its modern form, such an approach to teaching very rarely reflects what actually happened. For instance, a "history of communication" might lead back from the 1947 Bell Labs development by Shockley of the transistor (electronics). It used germanium, the deposits containing which were first located in the United States by nineteenth-century rockhound William Maclure (geology), who also funded a commune in New Harmony, Indiana, set up by Robert Owen, a mill-manager from Britain (textiles), who learned his utopian views from William Godwin, founder of the socialist movement (politics), whose daughter Mary wrote *Frankenstein* (novelist), after she married Percy Bysshe Shelley (poet).

The essays in this book follow the same kind of unexpected path in an effort to recreate a feeling of what it was like to be constrained by the contemporary context. I hope that each time the story rounds a corner the reader will experience a little of what it was like to be the characters themselves, and to be as surprised as they were at the turn of events. Sometimes the turn is a trivial matter, sometimes not. That's how things happened.

Earlier I mentioned the circular structure of these essays. There are two reasons why I make such play of the unstructured nature of history, but then, in this book, give it a formal shape. One reason is that otherwise these essays would have mirrored the serendipity I described, just going from anywhere to anywhere, with no reason for beginning where they start from, or ending where they go to (and leaving as many dangling readers as participles). Choosing to go round in circles, and to end each story where it begins, lets me illustrate perhaps the most intriguing aspect of serendipity at work, which shows itself in the way in which history generates the most extraordinary coincidences. In this sense, history repeats itself all the time.

You may not agree with the way these essays present events. That's fine. There is no single correct way to track from the past to the present. And if your disagreement goes so far as to drive you to find alternate routes for what I write about that are even better, write your own history. The more of us doing so, the better.

1

A BIT OF A FLUTTER

I SUPPOSE MY view of history tends away from the orderly and toward the chaotic, in the sense of that much overused phrase from chaos theory about the movement of a butterfly's wing in China causing storms on the other side of the world. Which is why I decided to have a go at reproducing the butterfly effect on the great web of knowledge across which I travel in these essays.

This thought came to me at the sight of a giant cabbage white in a *Lepidoptera* exhibit at the Natural History Museum in London, which reminded me of the other great Natural History Museum, the Smithsonian. Which owes its life to the persistence of one Robert Dale Owen. The two-term Democrat from Indiana almost single-handedly pushed through Congress the 1845 Bill accepting the Englishman James Smithson's $2-billion-and-change bequest (in today's money) that helped to set up the esteemed institution. Owen's efforts also involved unraveling one of the shadier deals in American financial history: most of Smithson's money, which had arrived in the United States a few years before, was at the time in the dubious grip of a foundering real estate bank in Arkansas, into which the U.S. Treasury had thoughtlessly placed it for safekeeping.

Owen was a liberal thinker, the son of a famous British reformer who had earlier started an unsuccessful utopian community in New Harmony, Indiana. Well ahead of his time, Owen championed women's rights, the use of plank roads (for rural areas not served by the railroads), emancipation, and family planning. This last he espoused in a pamphlet in 1830. Subtitled "A Brief and Plain Treatise on the Population Question" (which gives you a feel for the cut of his jib), it advocated birth control by everybody and included three examples of how you did it. Two years later much of Owen's text was lifted (unacknowledged) for a bestselling tract by Dr. Charles Knowlton of Boston: "The Fruits of Philosophy," which went into greater physiological detail.

Forty years on, Knowlton's/Owen's work was republished

by activist Annie Besant in England, where it was judged obscene and likely to pervert morals. Ms. Besant conducted her own defense at the trial and in doing so became the first woman to speak publicly about contraception. Which earned her a fine and a sentence. Undeterred, Besant took up larger causes: Indian independence (she was President of the first Indian National Congress), vegetarianism, and comparative religion. This was some years after she'd broken off a romantic interlude with another left-winger, a penniless nobody called George Bernard Shaw, with whom Annie played piano duets at the regular meetings of William Morris's Socialist League in London. Later, Shaw would become fairly well known as the author of *Pygmalion* and then world-famous as the author of its Hollywood remake, *My Fair Lady.* The play was all about talking proper (which Eliza Doolittle didn't, you may recall) and featured a prof. of elocution, Henry Higgins, whom Shaw modeled on a real-life linguistic academic named Henry Sweet.

In the 1880s Sweet was one of the inventors of the phonetic alphabet, interest in which was triggered by the contemporary craze for old languages kicked off by William Jones, a Welsh judge in Calcutta. In 1786 Jones had revealed the extraordinary similarities between the ancient Indian language of Sanskrit and Latin and Greek. The revelation revved up nationalism among early-nineteenth-century Romantic movement Germans (whose country had not long before lost a war with the French and was going through a period of cultural paranoia) because it gave them the idea that they might be able to trace their linguistic roots back into the Indo-European mists of time, thus proving they had a heritage at least as paleolithic as anybody in Paris.

This mania for reviving the nation's pride might have

been why German graduate students were also getting grants for such big-science projects as sending out forty thousand questionnaires to teachers all over the country asking them how the local dialect speakers pronounced the sentence "In winter the dry leaves fly through the air." On the basis of such fundamental research, pronunciation atlases were produced, and dialectology became respectable. So much so that at the University of Jena, a guy called Edward Schwann even got the money to do a phonometric study of zee French accent. Nice work if you can get it. Schwann was aided in his task by the eminent German physicist Ernst Pringsheim.

In 1876 Pringsheim was one of the science biggies visited by Franz Boll, a researcher who was working on the process by which the human eye is able to see in low light, thanks to the presence of a particular chemical. Or not, in the case of its absence. The whole view of such visual deficiency was taken a stage further by a sharp-eyed Dutch medical type, Christiaan Eijkman. This person happened to be in Java with a Dutch hospital unit, sent out there in 1886 to grapple with the problem of beriberi, a disease that was laying low large numbers of colonial administrators and army people. Eijkman happened to notice some chickens staggering about the hospital compound with symptoms not unlike those of the disease he was studying. But because these were chickens and not humans, he did nothing about it. Until suddenly, one day the chickens got instantly better. What kind of fowl play was going on here?

Turned out, the new cook at the hospital had decided that what was good enough for the local Javanese workers was good enough for birds. So he had stopped feeding to the chickens gourmet leftovers from the table of the European medical staff. Difference being in the rice. Europeans were given polished rice ("military rice"); locals and the chickens got the stuff with the hulls left on ("paddy"). Months of chicken-and-rice tests by Eijkman ended up with a meaningful thought: There had to be something in the rice hulls that was curing the chickens. Or, to put it *more* meaningfully, without this "something" in their diet, the chickens got the staggers. So was that why people did the same?

A few years later, in England, Gowland Hopkins, an ex–insurance broker turned biochemist, observed that baby rats wouldn't grow, no matter what they were fed, if their diet didn't include milk. He became convinced there was something in normal food that was essential for health and that wasn't protein, carbohydrate, fat, or salt. Gowland labeled these mystery materials "accessory food factors" and went on to share the Nobel with Eijkman, because their work would lead to the discovery of what these accessories actually were: vitamins (in the case of the chickens, thiamine).

Now, why all this made me think that how the web works might remind you distantly of chaos theory was because of what Gowland had been doing before he got into nutrition. He was able to work with pure proteins and their role in nutrition once new techniques had been developed (at Guy's Hospital in London, where Gowland had trained) to analyze uric acid proteins in urine.

And he was interested in uric acid because his very first scientific work had been with insects, when he had conjectured (wrongly, as it turned out) that uric acid was involved in producing the white pigment of the wings of the cabbage white butterfly.

2

SATISFIED CUSTOMERS

THE MODERN DEPARTMENT store, with its money-back-guaranteed merchandise, is one of the great examples of industrial democracy in action. Thanks to mass production and distribution, I can go back to the shop and get a free replacement copy for a cup that I found a flaw in last week. It was one of those willow-pattern things. Genuine Wedgwood. An ironic term, really, because Wedgwood's original stuff was fake. Josiah Wedgwood was a potter who started his career repairing Delft chinaware (fake porcelain, first made for the Dutch middle classes, who couldn't afford the sky-high prices of the real thing coming in from the Far East). Then, in 1769, Wedgwood graduated to crafting his own stuff (fake Greek vases, first made for the English middle classes, who couldn't afford the sky-high prices of the real thing coming in from southern Italy).

The source of Wedgwood's inspiration was an amateur archeologist and site-robber by the name of Sir William Hamilton, who had been appointed English minister to the court of Naples in 1764, not long after the first systematic excavation of the nearby ancient city of Pompeii. So there was a ton of classical bits and pieces lying around for what might charitably be referred to as "collecting." Hamilton's collection grew so big that he published catalogues, one of which influenced Wedgwood.

From time to time, Hamilton would return to England to sell his latest haul of antiquities to institutions like the British Museum or the duchess of Portland. On most of these occasions, the sales agent was his nephew, a ne'er-do-well called The Honourable Charles Greville. Now, there must have been something ne'er-do-well in the Hamilton blood, because Sir William's own mother had seduced the Prince of Wales, and in 1785 he himself took over Greville's mistress (to "save the boy the expense"). The lady in question was a strapping lass thirty-five years Hamilton's junior, who called herself Emma Lyon and who was into "attitudes" (posing, in diaphanous outfits, as various classical Greek and Roman personages).

Emma might have learned the trick while working as an

"attendant" for James Graham, one of the era's greater electricity quacks. Graham boasted an impeccable scientific background from Edinburgh University, where he had studied under such medical greats as Joseph Black, the discoverer of latent heat. Electricity at the time was something like cold fusion in the 1990s: Nobody quite understood it, but people supposed that it might do miracles. They knew that an electric current (produced by rubbing glass with a silk cloth, or by touching a Leyden jar) could cause dizziness, a quickening of the heart rate, and spots before the eyes. Maybe electricity was good for the health.

Graham claimed electricity cured only everything. At his posh London Temple of Health (in its elegant, Adam-designed premises), the elite took mudbaths and shocks while surrounded by scantily dressed nubile maidens (Emma was one for a while) and protected from the rude gaze of the riffraff by six-foot-tall bouncers on the front door. Graham had the London demimonde knocked out cold by the star of his shocking show: the amazing "magnetico-electrico-celestial" bed, guaranteed to fix infertility and almost anything else that ailed you. The line of the credulous infertile ran all the way around the block.

Back in Naples, Sir William Hamilton set Emma up in a plush villa, where she continued to assume attitudes. Not surprisingly, her posing turned out to be just the thing to catch the attention of a prominent Navy type who had been at sea for too long (that, and possibly the fact that, as he later noted, Emma never wore underwear). The sailor in question was the hero of the day, Horatio Nelson, whose charms were so renowned that when he sailed into Naples there was female fainting all round. He met Emma in 1798. Quicker than you could say "Admiral of the Fleet," she was his mistress, and they were canoodling on the island of

Malta, where the commissioner ruling the place was another old sea-dog, Captain Alexander Ball, who had once saved Nelson's ship and life.

In those days, Malta was a strategic hotspot in the conflict between Napoleon and the rest of Europe. Malta gave Nelson control of the Mediterranean sea lanes and hence secured the route through Egypt to British India. Which was why Napoleon was after Malta. And others. So the island was full of intrigue, and Russian, French, and Turkish spies. There were also a few Americans (resting up after their war with Tripolitania), who had their own transatlantic reasons for undermining the Brits.

All this international hugger-mugger meant that when Ball was not entertaining Nelson and Emma, he was busy writing secret dispatches, night and day. And, because Ball was better at navigation than prose, the dispatches were being edited, day and night, by his new rewrite man, a passing opium addict and Romantic poetry maven named Samuel Taylor Coleridge, who had arrived on the island in 1804, on the run from his wife and his habit.

Coleridge had journeyed to Malta to recover his health and financial well-being. After nearly two years, neither goal had been achieved, so the poet headed back to London via Rome, where he met and was painted by an American artist called Washington Allston. The two soon became close friends, and on a later visit to England, Allston introduced Coleridge to his protégé, a young American whose aim in life was to create one of the murals for the Capitol Rotunda in Washington, D.C. Alas, the job never came his way, although he did become the rage of New York's art world, founded the National Academy of Design, and made portraits of such movers and shakers as General Lafayette and DeWitt Clinton. In 1829, this young painter headed once again for Europe, where he gradually came to realize that his future might lie elsewhere than on canvas.

On the return trip, in 1832, he came up with the idea that made him so much more famous than did his art, that you are probably still wondering who it is we're talking about. The man was Samuel Morse, and the idea, of course, was sending messages along a wire. Six years of development later, Morse was only about the sixth guy to produce a telegraph, but his version hit the jackpot for at least two reasons. One was the Morse Code. Nobody is totally sure that he didn't snitch it from his partner (and supplier of free hardware) Alfred Vail. Be that as it may, compared with the complicated, telegraph-and-printer models developed by his competitors, Morse's technique was a breeze. It needed just a simple contact key (to send simple groups of five on-off signals), required only a single operator, worked over low-quality wire, and was cheap.

The other reason for Morse's success was also financial. Back then, railroads often ran both ways on single tracks (this saved money) and they frequently crashed (this lost money). Operators urgently had to find a way to instruct trains, coming in opposite directions, when to move and when to wait. The telegraph did just that, for the first time, in 1851 on the Erie Railroad. But it also complicated matters for its users.

By the mid-1850s the Erie employed over four thousand people and the rail network was growing like Topsy. In 1860, the company had around thirty thousand miles of track and things were threatening to go off the rails. The problem was that railroad companies served as many different enterprises all at once: shops, terminals, rail track, marshaling yards, warehouses, and engineering units. Moreover their materials, personnel, and money were spread across thousands of miles. And the nature of the business meant that, from time to time, they had to make instant, system-wide decisions. If the companies were to survive, they needed a radically new kind of command-and-control organization.

Three engineers came up with the solution, making use of the rapid communications facilitated by the new telegraph. Daniel McCallum (of the Erie), J. Edgar Thomson (of the Pennsylvania Railroad), and Alfred Fink (of the Louisville & Nashville Railroad) devised the first business administration organization chart, the idea of line-and-staff management and divisional company structure, and the first true cost-per-ton-mile financial analysis. As a result, the railroads were soon able routinely to handle thousands of articles (passengers and freight) at high rates of turnover (getting them on and off trains) at low margins (cheap prices) on a huge scale (all across the continent).

By the 1870s, railroad management techniques had helped establish another industry built on the frequent and regular delivery of goods. Like the railroads, these businesses operated on a large scale, at low margins, and with high-volume turnover. Like the railroads, their staffs outnumbered the population of many cities. And like the railroads, their organization was departmental. Which is why they became known as “department” stores. These places proved a great hit, and went on to generate the democracy of possessions that characterizes the modern industrial world.

So thanks in the first place to Wedgwood (whose factory is still operating), everybody today can buy his crockery. And anything else they desire. And if there is something wrong with it, get a free replacement, guaranteed.

A practice first introduced, in his London showrooms, by Wedgwood.

3
FOLIES DE GRANDEUR

SITTING HERE AT my trusty computer, I look out on the River Thames and Brunel's beautiful railway bridge, so I'm constantly reminded of the way nineteenth-century iron and steel technology gave them all machine-assisted *folies de grandeur.* So there I was, dredging my mental silt for a line on folly with which to start this essay, when one floated past, under the bridge. A dredger, that is.

Which suggested the Suez Canal, the *folie de grandeur* project of them all. Everybody, from the Romans on, had a go at it. Even Napoleon tried and gave up, when he invaded Egypt in 1798. His committee of scientists (who accompanied the troops) had told him the thirty-inch difference in water level between the Mediterranean and the Red Sea made it inadvisable. But in 1859, twenty-five thousand *felaheen* laborers, together with a financial consortium made up of Switzerland, Italy, Spain, Holland, and Denmark, finally succeeded. It was during the last stages of construction that suction dredgers were employed.

Both the canal and the pneumatic sand removal had been French ideas. The canal itself was masterminded by a think-big entrepreneur called Ferdinand de Lesseps (who then went on to bankruptcy over a similar job that didn't go so well in the Panama isthmus). Industrial-scale pneumatics had been introduced earlier, when the French were digging the first railway tunnel through the Alps under Mont Cenis. This was intended to unite Italian Savoy (north of the mountains) with the rest of Italy (south of them) and also make it possible for people sailing home from India and the East to pick up a train somewhere like Brindisi, instead of having to sail all the way round Spain. Unfortunately, before the tunnel was complete, war gave Savoy to the French. Still, the tunnel would be good for tourism.

In 1861, not far into the Alpine rock face (after three years of boring hand-boring, and advancing all of twenty centimeters a day), the chief engineer, Germain Sommelier, decided to try something that would finish the job in less than his lifetime—a specially built reservoir, high above the tunnel entrance, providing a head of water that would compress air to supply pneumatic drills that would get him through the rock faster. As it happened, the drills sped things up twenty times faster, but Sommelier never made it. He died of a heart attack a little later.

The Mont Cenis tunnel amazed everybody almost as much as the Suez Canal, and its new wonder-drills featured in a magazine picked up one day by a young American whiz kid, George Westinghouse. In 1869 he turned the pneumatic concept into an airbrake for use on trains. Compressed air, running through pipes underneath the train, held back pistons. In the event of the air pressure being released (either deliberately or by a rupture in the pipes), the pistons would slam forward, driving brake shoes against wheels. This would stop a 30-mph, 103-foot train in 500 feet, and made it possible to schedule more trains, more closely spaced than had previously been wise.

This in turn required better signaling. Which is why, in 1888, Westinghouse fell in with an inventive Croat who wore a new red-and-black tie every week and lived in a hotel room full of pigeons. Name of Nicola Tesla, this person came up with a way to send electrical power long-distance along railroad tracks, so as to operate railway signals. And then invented a little device so fundamental to the modern world that most of the time you don't even know it's there. He sent alternating current into two sets of coils wound on iron, setting up currents that were ninety degrees out of phase with each other. These generated a magnetic field that rotated with each successive burst of current. The rotating magnetic field caused a copper disc to spin. When you put a belt on the disc you had an electric motor.

By World War I this trick was just what the captains of the new monster-size battleships were looking for. First of all, because metal ships carrying on-board electric power had been making life difficult for a magnetic compass. So you could easily get lost. And second of all, in a rough sea the giant new fourteen-inch guns, which fired 850-pound shells nearly ten miles, couldn't hit an enemy barn door if the ship were rolling heavily at the time. Tesla's little motor helped solve both these problems, because it could spin gyroscopes of at least three different sizes. There were tiny gyros for true-north-pointing gyrocompasses (once you spin the gyro, if you leave it alone it stays pointing the way you set it spinning, come hell or, more appositely, high water). There were also humungous, four-thousand-ton gyros, spinning in the center of a ship and compensating for the roll of the sea. And finally there were midsize gyros, doing the same favor for all the gun platforms and permitting dreadnoughts to live up to their name. In her first wartime encounter, the newly gyro-stabilized USS *Delaware* shot every attacking plane out of the sky. During a storm.

This use of the gyroscope was the brainchild of a Brooklyn electrical component manufacturer called Elmer A. Sperry, and it made his fortune. Mind you, persuading the Navy to buy hadn't all been plain sailing. And the financial risks were real high-wire stuff. Which, as it happened, was where Sperry had originally (and for only once in his life) failed. Early on, he'd tried to talk showman P. T. Barnum into featuring a gyro-stablized wheelbarrow in one of his circus trapeze acts.

The likely reason for Barnum's refusal was that he didn't have much to do with technology except briefly, in the 1840s, when he first set out to be a showman and went looking for curiosities to exhibit. His wish list included "industrious fleas . . . fat boys . . . rope dancers . . . and knitting machines." Moreover, by the time Sperry was pitching the gyro idea, Barnum was well beyond wheelbarrows (or industrious fleas, for that matter), touring "The Greatest Show on Earth" (eight hundred people, ten thousand miles a year by special train), and had already invented the three-ring circus. Known as the "Prince of Humbug," Barnum could have sold refrigerators to Eskimos. In his Big Top, the crowned heads of Europe went crazy over his midget admirals, buffalo hunts, and elephants, as well as spectacular reruns of the such modest moments as the Destruction of Rome, the Fall of Babylon and the American War of Independence.

From time to time Barnum's personality would do a flip-flop and he'd give it all up for temperance work. Or, on one occasion, in 1850, to manage a U.S.–Cuba tour for the greatest soprano in the world, Jenny Lind. In 1844 Ms. Lind had given her first performance outside Sweden (in Berlin) and was so extravagantly successful that she became an instant diva at the age of twenty-four. People paid crazy prices for tickets to her appearances, even if they couldn't make it. One fan wanted only to touch her shoulder "to see where the wings began." Queen Victoria threw her own bouquet at Lind's feet. In the street she caused scenes that wouldn't be witnessed again until the Beatles. In 1845 Her Majesty's Theater in London decided to commission a new opera for Lind and offered the job to the other contemporary operatic superstar, Giuseppe Verdi. Two years later Verdi obliged with *I Masnadieri*, starring Lind in the role of Amalia. Boffo success.

Verdi was always delighted to accept foreign writing

work because it paid up to seven times more than he got at La Scala, Milan, in spite of the fact that he was Italy's premier musical nationalist at the time. In the 1840s Italy was occupied by the Austrians, and Verdi was craftily getting round problems with the censor with inflammatory stuff about killing *Swedish* kings, and going on about *Israelites* in bondage, and *American* revolutionaries, and other "geddit?" subtleties like that.

This might be the reason Verdi got the chance to write what turned out to be the most popular opera ever: *Aida*. Ruling Egypt at the time was a khedive (ruler) named Ismail, whose local engineering improvements had cost so much he was severely short of funds, and had to sell his shares in an engineering project that he had hoped would be a major national money-earner (and to celebrate which *Aida* was originally commissioned). The opera, set in pharaonic Egypt, was also supposed to glorify the country's ancient past and cock a snook at Ismail's Turkish overlords.

It didn't do much in that particular vein, perhaps because it took so long to work out a deal Verdi would accept that delivery of the score was nearly two years too late for the occasion it was supposed to celebrate: the opening of the Suez Canal.

Which ends my machine-assisted *folie* for now.

4

A LOT OF BALONEY

I HAVE TO confess a fatal weakness for Bologna, Italy. Apart from the fact that it has the oldest university in Europe, and the most elegant women on the planet, it also happens to be the food capital of the known universe.

After visiting the Diana restaurant (unsolicited recommendation) and lunching on works of gastronomic genius (don't miss the tortellini alla panna), you can walk a few hundred meters and savor another work of mouth-watering precision: a giant brass meridian line, inlaid across the floor of the city's cathedral. Put there by Gian-Domenico Cassini, the hottest astronomer around in 1667. Which was when his reputation brought him an offer he couldn't refuse from Louis XIV's right-hand man, Jean-Baptiste Colbert: to run the new Paris Observatory. And then get involved in the great French effort to identify the shape of the Earth (which the French thought was not flattened at the poles).

Colbert needed to know such arcana so that the new navy he was building for France would more accurately be able to relate star-fix angle to position on the planetary surface (different on an Earth that *was*, or was *not*, flattened at the poles). This way, French ships would be able to navigate better. And rule the waves. And also, perhaps, give the English one in the eye, by snitching the Prime Meridian from Greenwich and moving it to Paris. Unfortunately for French *amour propre*, they were wrong about the shape, which is why I'm writing this in GMT.

The sidereal shenanigans were a key element in Colbert's grand plan to make France a mercantile superpower, as part of which he also offered tax breaks to anybody interested in sailing off (more accurately, it was now hoped) to exotic foreign parts and coming back with import deals for high-end consumables. Idea being, Colbert could then turn this trade into a French monopoly and make oodles of ecus for king and country. Well, king.

This perfectly legitimate scheme for avoiding tax was yet another offer too good to refuse, so in no time at all, freebooters were returning from Senegal in West Africa with shiploads of gold, ivory, slaves, and gum. Senegal gum turned out to be just what you need to machine-print chintz (the newest Euro-fashion-craze from India) with fast colors, because the gum acts as a dye-binding agent. By mid–eighteenth century, the chintziest guy in town was an Irishman named Francis Nixon, who invented a method for producing all the cheap prints you wanted, at low cost and high speed. Nixon's trick involved pressing steel designs onto copper rollers, then coloring up the rollers and running cotton between them. Thus inventing matching curtains and covers.

Anyway, by 1818 young Jacob Perkins of Newburyport, Massachusetts, had added a few minor modifications to Nixon's process and was in London trying to persuade the Bank of England to give him the printing contract for banknotes whose designs he claimed were so complex nobody could fake them. Speedily (eighteen years later) the Bank said yes. Well, this *was* England. Four years after that, the patient Perkins really licked his printing competitors when he landed the job of producing the new British "Penny Black," the world's first postage stamp.

Businessmen all over the place got instantly stuck on this exciting new idea of pushing the communications envelope (OK: sorry!). By 1874, thanks to the introduction of the stamp to Switzerland by the world's leading banana expert (more of him in another essay), Berne was home to the Universal Postal Union, and the international community was deciding to divide postal materials into three categories: letters, parcels, and a brand-new thing called a "postcard."

Illustrations on some of the earliest cards came from the talented pen of Phil May, a British cartoonist who did his best work for a new satirical magazine called *Punch*. The magazine hadn't planned to include drawings, but events presented one of those opportunities every editor dreams of.

When Queen Victoria's husband, Prince Albert, came up with a competition to decide who should fresco the interiors of the newly rebuilt in nineteenth-century imitation Gothic Houses of Parliament, the competition entries that His Royal Highness liked best were so ludicrously *dreadful*, there was only one way to stop him. Publish them. It worked.

But not even Punch could prevent the Gothic Revival. Next time you're in Britain, note how many nineteenth-century churches there are. You'll know them from real Gothic because when they were built they featured both gargoyles and gaslight. Gothic was cheaper than Neoclassical, so the Victorian church commissioners threw up more than five hundred of these piles. All the fault, I suppose, of that late-eighteenth-century back-to-the-Middle-Ages lunacy we now call "Romanticism," spearheaded (like that medieval touch?) by a young German philosopher named J. G. Herder, who was deeply into the fundamental unity of humankind, nature, German folk songs, and something called *sturm und drang*—an epic view of existence, possibly best translated into modern parlance by the phrase "over the top."

Herder had been turned on to such excesses by the clamorous arrival in Germany in the late eighteenth century of a collection of ancient Gaelic poems written by the third-century Irish warrior-poet Ossian. These took Europe and Herder by storm (OK: sorry!). For Romantics, the epic pulsated with the pure and powerful feelings of a primitive people. And thanks to what it did to Herder and his pals, triggered the Romantic Movement. Of such things are great moments of history made. Too bad the epic was fake. "Discovered" by a middle-rate Scots poet called James McPherson, who bundled a few ballads he'd collected (on a tour of Scotland) together with his own work, translated into Gaelic, and passed off as a fifteen-hundred-year-old original. Still, as

I said, he did help to give us Romanticism, which in turn gave us pathology and radio (of which more in another essay).

Meanwhile, why would a guy like McPherson be on an antiquarian ballad-hunt in the first place? Well, I suppose because back then, the future for Scottish culture was looking bleak. As a result of the fact that, ever since 1715, the Catholic Stuarts had been submitting claims to the English throne (now occupied by Protestant Germans) in the form of armed uprisings, so there were Redcoats all over the Highlands. Where matters such as clans, tartans, and speaking the local lingo were all things that could get you seriously hanged. The English even wrote a special extra verse to the National Anthem, all about "rebellious Scots to crush!" Things came to a head in 1745 when the last royal Stuart, Bonnie Prince Charlie, and his murderous band of cutthroats (aka "band of brave patriots") got as far south as Derby, in consequence of which there was a run on the pound. Now, you don't mess with the Bank of England and get away with it. But get away he did. In the words of the song, "over the sea to Skye," and then across the Channel to the Continent. All thanks to a supporter named Flora McDonald (who subsequently lit out for North Carolina). To this day, in memory of Charlie's flight to foreign parts, romantic Scots will raise their glasses to "the king over the water."

And finally: the reason why I began this particular example of tortuously crafted baloney the way I did. Because . . . guess where Charlie ended up spending the best of his declining years in exile? Well, where would *you* go, if (like him) you were looking for intellectual chat, enjoyed the company of elegant women, and overindulged in food and drink (of which the Prince ultimately expired)?

Whichever way you sliced it (OK: sorry!), there was only one choice: Bologna.

5

IMPRESSIONS

FORTUNATELY FOR ME, the other evening at a reception to mark the opening of an art exhibition I noticed a woman drinking a glass of champagne and got the impression she was closely scrutinizing one of those French paintings you can really only appreciate from a distance. I say "fortunately," because the event provided me with an idea for this piece (and several glasses of champagne).

Early in the nineteenth century, up to his ears in conflict with practically everybody else in Europe, Napoleon must have been thoroughly fed up with the fact that thanks to the massive levels of industrial output by the enemy Brits, he was fighting them armed with British-built cannon manned by troops wearing uniforms made in England! *Zut!* So he set up a Society to Encourage French Inventors (very rough translation), and in 1810 a total nobody named Nicholas Appert stepped forward to take the Society's prize of twelve thousand francs for a crazy idea he'd tried out on the French navy a year or so earlier. Appert had come up with an idea for preserving food. All you had to do was seal the food in a champagne bottle (Appert was a cook and champagne bottler), and then immerse the bottle in water, which you then brought to the boil for long enough to kill the germs that caused putrefaction. As is so often the case with these major advances in science and technology, Appert didn't know that bactericide was what he was actually doing. But never mind.

Poetic ravings about about how M. Appert's bottled veggies "brought spring and summer to winter" appeared in the French press, at which point the Brits got to hear about it. In wartime 1811 an Anglo-French go-between, John Gamble, one of the British prisoner-of-war exchange team in Paris (who was also married to a Frenchwoman), managed to get hold of Appert's patent. One year later, together with a couple of partners (Bryan Donkin and John Hall), Gamble set up a business in Bermondsey, South London, repeating the food-preservation trick, but this time in tin cans (one of his partners had experience in iron-making).

After the British royal family had tried some of the new products and pronounced them "delicious," how could canning fail? In 1818 the industry got another major boost when the exploratory Captain John Ross sailed off in a blaze of publicity to find the Northwest Passage, carrying a large supply of the canners' carrots and gravy, soup, roast veal, and peas.

In 1824, the intrepid captain's next, similarly provisioned expedition (funded, in keeping with this gastronomically oriented article, by Felix Booth, distiller of the eponymous gin) discovered the Magnetic North Pole. And named a very northern bit of North America the "Boothia" Peninsula. The actual magnetic discovery was made by Ross's nephew and traveling companion, James, who was so bitten by the polar bug that in 1839 he shot off in the opposite direction, on board HMS *Erebus*, to spend four years finding and mapping large bits of Antarctica and other spots en route.

On this occasion, one member of his crew was a young man called Joseph Hooker, who afterward became famous by writing up the botanical discoveries made on the trip, and then going on to do the same thing again on assorted sorties to Sikkim, Nepal, Assam, and India. As a result of these Himalayan ramblings, Hooker became known to gardeners everywhere when he introduced the West to most varieties of rhododendron, and then over a number of years patiently catalogued more than three hundred types of impatiens. For such persistence, in 1865 Hooker was made director of the British Royal Botanical Gardens at Kew (following his father in the job) and proceeded to turn the place into the international center for botanical research it is today. He also saved many a latter-day tourist (and me) from the rigors of bone-chilling London winter afternoons, when he commissioned the tropically warm splendors of Kew's beautiful Palm House. Speaking of which, Hooker did at least two other things that matter to the twentieth century. He helped to organize the smuggling of rubber seedlings out of Brazil (not British at the time) so that they could be nurtured and then transplanted to the Malaysian archipelago (mostly British at the time), thus laying the foundations of the entire rubber industry, and making possible the invention of the raincoat.

Hooker went on to do the same trick for the West African oil palm. About which you should only concern yourself if you happen to be watching your weight and laying off the animal-fat intake. Palm oil really came into its own thanks to Napoleon's nephew (Napoleon III) and *his* problem: feeding the troops and a rapidly rising population. In response to yet another imperial call to the flag (and the offer of another fat prize), a French chemist called Hyppolyte Meges-Mourriès changed the nature of the sandwich with what was, in its earliest form, a mixture of animal fat churned with milk and salt, chilled, kneaded, and packaged. Alas, poor old Hyppolyte never got his hands on the prize money. And to add insult to injury, certain others, recognizing on which side their financial bread was buttered (so to speak), promptly took advantage of patent-law loopholes to mass-produce their own versions of his new food substitute (known as margarine), and to become major modern industrial giants (in later years, using palm oil in preference to animal fats).

Meges-Mourriès had derived all he knew about fats (and probably also the name he gave his invention) from the great Michel-Eugène Chevreul. In 1889, when Chevreul died at the age of 102, France declared a day of national mourning, because Chevreul's research into fats and oils had made the world a brighter, sweeter place: He'd turned soap-making into an exact science, and he'd invented a better candle. He'd also done some good things for French tapestry-making. In 1824 he'd been made director of dyeing at the great Gobelins factory (because the way organic dyes act in fabric has a lot to do with plant oils, and he was hot stuff on such things). As part of his work on color (his word "margarine" comes from the Greek for "pearl-colored"), Chevreul got turned on by the way hues are perceived, and produced his "Law of Simultaneous Contrast." This said that the way a color is seen has to do with whatever other colors are placed next to it. Such might well have been an observation the Gobelins weavers had made with their very first throw of the shuttle, but, as far as I know, nobody had yet looked at the matter scientifically.

There was only one bunch (apart from the weavers) who cared deeply about this color-juxtaposition thing Chevreul had discovered: a guy called Georges Seurat and his painter pals, excited by what you could do with a lot of little dabs of different color placed in close proximity to one another. Which is, I suppose, an offensively over-simple way to describe what the art world knows as "pointillism." Demonstrated brilliantly, in 1886, by Seurat in his *un Dimanche d'été à la Grande Jatte*, one of the greater works of the so-called "scientific" Neo-Impressionist school he founded. Another example of which was being examined by that woman I mentioned at the beginning (in the champagne reception to open the exhibition, remember?).

One last little touch. Guess where Seurat's family came from. Champagne.

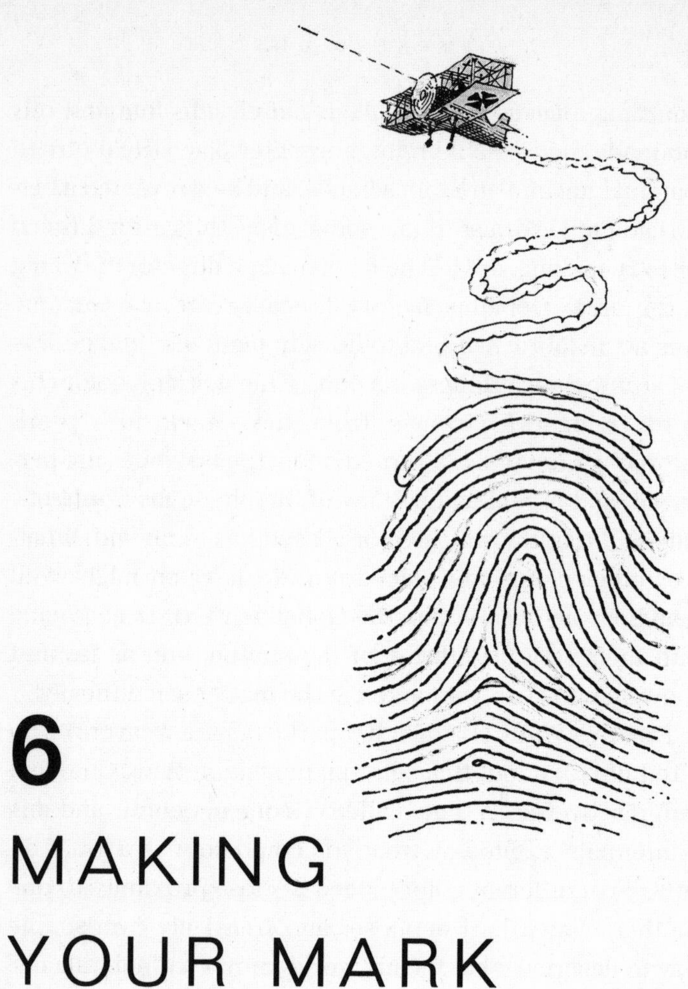

6 MAKING YOUR MARK

I WAS WATCHING the news the other night when I saw a story about somebody being identified by the so-called "DNA fingerprinting" technique, and there on the screen were the now-familiar black and white bands you're starting to see in all the cops-and-robbers movies. All thanks to a Swedish Nobel laureate, Arne Tiselius, who made all this fun possible back in the thirties when he worked out how to make protein molecules line up according to their weight, by putting them in a gel and then zapping it with various charges. The heavier the molecules were, the less far they moved. Electrophoresis.

The hardly visible differences in the gel caused by the presence of the proteins was more easily seen with *schlieren* photography, which shows up the slightest change in density because of the way light behaves when it comes through differing media. Like gel containing different amounts of protein. Or like turbulent air. Which is why, right from the start, *schlieren* was also a great success with airflow freaks like Theodore Von Karman (whose famous vortices you'll sometimes see at takeoff on a damp day, curling away from the wings).

Both Von K. and his curls were well-known to a pal, Anthony Fokker, of the planes of the same name. But Fokker did something else besides build great flying machines that would first cross the United States nonstop in 1926, and a year later take Byrd over the North Pole for the first time. During World War I the Germans captured a French invention, and Fokker turned it into a clever way to synchronize a propeller with a machine gun, so fighter aces wouldn't shoot their props off. And since all you had to do with this new system was point the plane and fire, the arrangement was so successful (in the hands of hotshots like Manfred von Richthofen, the Red Baron, in his red Fokker) it became known as "the Fokker scourge."

At the time, machine guns were doing well elsewhere, too. Specially when thousands of infantry were snagged on barbed wire during an advance. Sitting ducks, really.

Which is what the amazing new tank was supposed to prevent, by crawling over (and knocking down) the barbed wire, thus clearing a path for the troops. Irony was, that the tanks were driven by members of the old cavalry regiments, because it had been lack of horse in wartime America that had kicked off traveling armor in the first place. Most of America's farm horses had been requisitioned for use by transport regiments in Europe, to carry supplies to the men who were now *also* missing from the farms. Back home, this shortage of horse-and-man-power got a fellow named Ben Holt inspired to invent an entirely new kind of agricultural implement, because in the San Joaquin Valley (where he was at the time) a lot of the land was so wet it wouldn't support anything on wheels (or in some cases, hooves). Holt invented a crawler tractor, running on tracks that spread the load. A friend of his remarked it looked like a caterpillar, and the rest is farm-equipment history.

Most of the early Holts were sold to the Allies in World War I, were noticed by the military, and bingo: tanks. Holt's early tractors ran on gasoline, but later he switched to (and made a great success of) the diesel engine, which was itself so successful because it was cheaper to run than gasoline engines, it would start cold, and people thought it would burn almost any junk (there was even one version that ran on peanut oil). It may have been this last selling point that got Diesel his financial backing in the first place. Because right from the start, when talking to anybody about his engine he used the magic word "coal."

It was how the engine worked that made this possible. With a diesel, all you do is compress air in a cylinder so its temperature goes up to near eight hundred degrees Centigrade. At the top of the piston stroke, when the air is hottest, you inject a suitable liquid, gas, or particulate. The heat of the air causes it to combust. That pushes the piston down, to start the cycle all over again. So the diesel looked as if it'd burn anything that . . . well . . . burned. Like coal dust. In 1897, this sounded like sweet music to a man who ran the biggest steel-making plant in Europe, ran several railways, and owned all the coal mines to supply their fuel: Fritz Krupp.

Not surprisingly, Fritz's father, Alfred (who'd brought the firm to greatness earlier in the century), had been less than keen on the new wind of socialism sweeping through the industrial world at the time. For him it meant anarchy and the end of order. So he came up with a way to keep his workers sufficiently happy not to want such revolutionary stuff as trades unions. He gave his workers canteens, pensions, housing, company discount stores, and even uniforms to wear at home (what would you expect from a guy who once said: "As pants the deer for cooling streams, so do I for regulation").

It was this welfare scheme (as well as his love of trees, and dislike of company) that bonded him with the fellow running Prussia at the time. This person was, I suppose, regulation personified. Otto von Bismarck (who also loved trees and hated company) and Krupp were meant for each other. After all, Otto made war and Alfred made guns. Bismarck was also keen on welfare (he started the first universal old-age pension) and statistics and such, because the more numbers you had on all the average joes out there in the streets, the better you could regulate people so as to increase national output (or fight wars). Bismarck had developed this passion for the average joe because Ernst En-gel, the head of the Prussian Statistical Bureau, greatly admired the Belgian astronomer who had *invented* the average joe: Alphonse Quételet.

In 1835 Brussels, Quételet modified the kind of math that astronomers used (to calculate the probable path of heavenly bodies of which too few sightings had been made to be certain) to do the same thing to population statistics, because he believed if you applied this math to large numbers of people you could develop what he called "social physics." This way, you could work out what the average joe was up to, and do statistically meaningful sampling. Well, this was better than previous counting methods (like multiplying the number of chimneys by an assumed average family size to get the population), so it attracted the interest of such mathematical biggies as Charles Babbage.

His involvement with Quételet led to the formation of statistical societies in Britain and eventually inspired a young man called Francis Galton to go looking for ways in which any individual could be singled out from the mass. The result of his work was the discovery of a sure-fire way to spot one person from another, known as the fingerprint. Which was, until the advent of the DNA version I saw on TV the other night, the ultimate ID.

PS: Ironically (given how valuable electrophoresis was to prove in the original discovery of DNA), guess who Galton's cousin was. Darwin.

7

WHAT GOES AROUND COMES AROUND

BACK IN THE dawn of time I reported for the BBC whenever an Apollo lifted off, so nowadays every time a Shuttle comes back for recycling I am reminded of those long-gone, "use-once-and-throw-away" Saturn V days. On one of which, in 1973, a crew went up to the orbiting Skylab (itself a piece of converted leftover Apollo hardware), where major discoveries were then made about the solar corona having dark bits in it, which meant significant things to those interested in solar wind. And why not.

Of course, when you think Apollo, you can't not think Werner von Braun, can you? The German engineer who made all those extraordinary Saturn launch vehicles possible because (as he himself said) he'd had lots of earlier practice with all the wartime missiles (aka "Vengance Weapons") that left Peenemunde at regular intervals and extremely high speed, heading for England (and me). Devilish difficult things to hit, buzz bombs. At first it took the antiaircraft gunners twenty-five hundred rounds each to bring one down. Then (and on behalf of the U.K. I thank you, Bell Labs) along came the M-9 predictor, and the average cost of bagging a doodlebug went down to hundred shells. Significant savings all round (including that of life). All it took was the magic math to run a feedback loop constantly updating the set of data you got from the radar about the target's *last* position, so as to be able to make a pretty good guess at its *next*, and tell the guns to point that way. And then: boom.

The whole idea of feedback came to those involved with guns and such from those involved with gastric juices and such. Physiology had been looking at the body's feedback internal balancing systems (you know: hot-sweat, thirst-drink) ever since a mid-nineteenth-century Beaujolais wine-grower medic called Claude Bernard first noticed something odd about rabbits. In simple terms, if they were given no food for a while, the little bunnies appeared to live off themselves (we would say their own reserves of fat), and the difference in diet showed up in the turbidity of their urine. Which led Bernard to further discoveries like how the liver secreted exactly the amount of sugar you happened to be short of (or not, if not). And then all the rest of those reactions that happen in order to maintain what we now call homeostasis.

Mind you, there were those around who were far from happy about the rabbits. And the dogs, frogs, ducks, guinea pigs, and rats on which Bernard and others performed their various tests. Some of which, in all honesty, might not have been strictly unavoidable. (Bernard said: "The science of life is like a superb salon, resplendent with light, which one can enter only through a long and ghastly kitchen.") It was all too much for Bernard's wife, who left him and joined the antivivisectionists, one of whom was an Englishwoman called Anna Kingsford who spent most of her time in Paris, directing all her energies into killing Claude Bernard (and other high-profile physiologists like Paul "the bends" Bert) with thought waves.

The antivivisection group eventually became known, mistakenly, as the Humane Society movement. In fact, Humane Societies had been around since the eighteenth century, and they weren't there to save animals at all, only the "Apparently Drowned," all of whom were human. Many of these unfortunates had, for various dubious reasons, simply fallen into water. But there was another, less questionable motive for being considered Apparently Drowned. All you had to do to qualify was jump into the raging seas when your ship was on the rocks (so there was nowhere else to go but over the side), and then, with a bit of luck, be rescued by an offshoot of the Humane Society named (from 1824) the Shipwreck Institution.

Now the reason for all its cork-lined boats and crews was that by this time there were many more ships heading for trouble, rather than for port. What with all the extra cargo floating across the oceans thanks to the Industrial Revolution (with tons of manufactured goods going one way and raw materials the other), there was now more money being lost than ever before when people's ships didn't come in. Hence all the fuss and cork.

All the more so after midcentury, when an American naval officer named Matthew Maury persuaded hundreds of ship's navigators to keep (and send him) regular logs of wind, pressure, temperature, and currents. He also persuaded them to throw daily bottles overboard (sealed), carrying bits of paper on which they had scribbled their latitude, longitude, and the date. Said bottles to be picked up by others doing the same. Result: a zillion details on wind and current conditions that led to Maury's famous publication: *Sailing Directions.* These illustrated what might be called oceanic expressways, and indicated the fastest (cheapest) route from any A to any B. Which was another good reason to set sail for foreign parts with your cargo. And, as I said, very likely sink.

One of the other sources of Maury's knowledge of wind and water was the French physicist J.B.L. Foucault, who some years earlier had invented an amazing pendulum, which hung on a very long wire and proved that the Earth turned. As the (inertial) pendulum swings back and forth, it appears with each swing to trace a changing path (clockwise, in the northern hemisphere, and down south, vice versa). What this told people like Maury was that as the planetary surface rotates, in the northern hemisphere the path of winds coming up from the equator would be deflected to the east (and down south, vice versa).

Foucault's inertial investigations also led him to other matters affected by the turning planet. Problem being: Life for astronomers wasn't made any easier by the fact that, thanks to the Earth's rotation, they and their telescopes were whizzing around at a fair lick (no observatorial pun intended), making it hard to keep an eye on stars and stuff (which, after all, were not doing the same thing). So Foucault came up with a clockwork gizmo (well, siderostat) that rotated a mirror through 360 degrees, in the opposite direction to the Earth, once in twenty-four hours. This kept the mirror constantly pointing at any heavenly target, whose reflection could now be examined without the usual panic. Which made it easier for Foucault to do his next trick, in 1845. This was to use the new Daguerreotype photographic apparatus to do portraits of various heavenly bodies, now that they were sitting in the frame long enough for you to take their picture. This nifty process went over very big with stargazers everywhere, because of the many things you can do with photographs of starfields (like count the stars, or overexpose the plate and see many more).

In Koningsberg, Germany, during the 1851 eclipse a professional photographer named Berkowski used the same technique to take the first photograph of the phenomenon that the Skylab crew would study, from orbit, 122 years later: the solar corona.

Proof that in the world of connections (and orbits) what goes around comes around.

8
SWEET DREAMS

ONE OF THE less glamorous aspects of my work is having to fly frequent red-eyes, and any airline that offers me a sleep-inducing hot chocolate gets my money.

So there I was the other night, droning up into the night

sky, sipping, and thinking about Dr. Sir Hans Sloane. This eighteenth-century polymath helped to establish the place where I do most of my research, when his collection of twenty-five hundred plants, animals, and assorted memorabilia became the core of what would end up as the British Museum. And it was Sloane (while spending time in 1688 as personal physician to the governor of Jamaica) who discovered the beneficial and soporific effects of mixing chocolate and hot milk.

Back in England, in 1715 he treated one of the great literary wits and beauties of the day, Lady Mary Wortley Montagu, who was suffering from smallpox (she ended up with no eyelashes and a pitted face). The following year, when Lady Mary moved to Turkey with her incompetent ambassador husband and saw what the locals there were doing for smallpox (inoculation), she carried out Turkish-style treatment on her own son and then came back to England and got various royals to inoculate their kids. Then, with the active help of Sloane, everybody else did, too.

In 1736 Lady Mary found herself fatally attracted (as were various other lords and ladies) to an androgynous Italian science type named Francesco Algarotti, half her age, who was visiting London and working on a rewrite of Isaac Newton's work for women readers. Francesco and Mary fell hopelessly in love. Well, *she* did. Three years of one-sided passion later, Lady M. headed off for Italy and a change of air, after arranging a secret rendezvous with Algarotti en route. Of course he never showed, having gone off to Prussia to be court chamberlain (the Prussian crown prince Frederick had also apparently fallen for him, to judge by the fact that he went around referring to Algarotti as the "Swan of Padua").

Algarotti was a bit of a social climber (you noticed), so he must have been tickled pink when his interest in Newton got him a social invitation as rare as hen's teeth. It was to stay at the château in Champagne where the extremely reclusive philosopher François-Marie Arouet—aka Voltaire—was holed up (the French state security cops were after him for misdemeanors such as saying England had a better political system than France) together with his intellectual paramour Emilie, marquise de Chatelet. Each of them was also working on a version of Newton (he, a general book for lay readers; she, a rather more demanding commentary on the *Principia Mathematica*).

Voltaire also happened to be a great admirer of a pal of Algarotti's, an Italian priest-experimenter called Lazzaro Spallanzani, who (a hundred years before Pasteur) noticed that putrefaction didn't occur in hermetically sealed vessels. He also investigated how flat stones skipped on water, climbed into volcanoes, rode rafts into whirlpools, and sliced up thousands of worms, snails, salamanders, and tadpoles to test their regenerative abilities. He reported the first-ever case of artificial insemination of a spaniel. Above all, Spallanzani made real enemies among the scientific establishment by rubbishing the popular idea of spontaneous generation of life (that maggots came from decaying meat and mice from rotting cheese, and so forth). Spallanzani traveled so far and wide and became so famous that he inspired the wizard-genius character in one of a collection of creepy pseudoscientific yarns (full of maniacs and automata and ghosts and such) written around 1815 by a Berlin lawyer, and known, from the author's name, as the *Tales of Hoffmann*.

Apart from having his story lines snitched by such operatic personages as Tchaikovsky, Delibes, Offenbach, and Wagner, E.T.M. Hoffmann's other main claim to fame was that he defended a pan-German liberal fanatic named Friedrich Jahn. The recent defeat by the French had brought German university student mobs into the streets calling for a united Germany, free speech, democracy, and other such dangerous lunacy, and Jahn was in court because much of this dangerous ranting and raving was his idea. That, and gymnastics. Which Jahn saw as the only way to make German youth strong enough and disciplined enough for "the struggle ahead." A crackdown on Jahn's adherents followed his trial, on the official grounds that gymnastics might be detrimental to state security.

After one of Jahn's followers carried out the spectacular 1819 stabbing of a well-known establishment figure, August Kotzebue, feeedom of the press was abolished, all universities were taken over by the state, and another Jahn disciple, the athletic liberal Karl Follen (a suspiciously close friend of the Kotzebue stabber) fled to Harvard. Where he opened the first college gym in America and started everybody jogging. By 1851 German-American gymnastics was so widespread that a hundred gym clubs got together as the Socialist Gymnasts' League and were chosen to provide Lincoln's bodyguard at his inauguration in 1861. German gymnastics really took off in the United States when the YMCA adopted physical health as one of its basic tenets and built some of the first public gyms (and eventually invented basketball to play in them).

The international nature of the YMCA movement had already been established as early as 1855, with a world conference held in Paris. This get-together was the idea of a Swiss libertarian and evangelistic Christian called Jean-

Henri Dunant, who was the main author of the YMCA Charter. In 1859 Dunant happened to be in Italy watching the one-day Battle of Solferino (people used to take picnics onto hills next to battlefields and enjoy the carnage). Dunant was so appalled at the condition of the six thousand wounded that he rounded up three hundred locals and battle-watching tourists, who then used buckets of water to wash wounds all night. In 1862 his book *Memories of Solferino* led to the founding of the Red Cross. By the end of the century the Red Cross was on every battlefield, doing everything but blood transfusions (probably the one thing the wounded needed above all else, but which, when attempted, often seemed to kill rather than cure).

It took a Viennese immunologist named Karl Landsteiner to make transfusions possible when, in 1909, he discovered the four main blood groups A, B, O, and AB, and in doing so revolutionized surgery at a stroke of the hypodermic. In 1922 Landsteiner moved to the Rockefeller Institute for Medical Research in New York, and in 1930 he collected his Nobel for his blood-group work.

That same year, in the same institute, another Nobel winner (and needle expert), Alexis Carrel, who was a French immigrant surgeon and who had developed new suturing techniques that changed the world of blood vessel surgery, took a major step forward in organ transplant work (for which Landsteiner's blood-matching discovery was essential). The success followed Carrel's collaboration with a chap whose sister-in-law had a heart valve problem, which was inoperable because at the time there was no way to maintain the patient's circulation during the operation. After somebody introduced this guy to Carrel, they were a dream team. Over the next few years the new man developed a new perfusion pump for Carrel that used compressed gas to maintain continuous circulation of the necessary fluids. In 1938 both men made the cover of *Time* magazine.

The pump-maker was used to the publicity, of course, because he was already world-famous. He was Charles Lindbergh, in 1929 the first person to fly the Atlantic and the man who made possible flights like the one on which I nodded off at the beginning of this essay.

Hope you didn't.

9
WAVING THE FLAG

I WAS LOOKING up at the Stars and Stripes during a ceremony a few weeks back in Washington, D.C., and thinking about how things might have turned out if Lord North hadn't been away for the weekend when the American Revolution (sorry: War of Independence) happened. There would have been no difference, probably, since French financial backing had made absolutely certain, long before, that the United States would end up the United States, mainly because if there was anything the French wanted (more than a good meal and a great bottle), it was to cock a snook at the Brits. Some snook.

The whole transatlantic adventure was masterminded by a multitalented guy named Caron de Beaumarchais, watchmaker to Madame de Pompadour, and the most hard-done-by writer for the French theater (he wrote *The Barber of Seville* and *The Marriage of Figaro*, snitched respectively by Rossini and Mozart). In his spymaster role Beaumarchais was the CIA before the CIA. In 1775 he went on a secret mission to London and reported back that the British grip on America was weakening and Louis XV would be well placed to do something to loosen it further. So Louis put Beaumarchais in charge of the whole scam. This involved setting up a fake company (Roderigue, Hortales and Company) to launder French funds to the American rebels (sorry: patriots) and chartering a fleet to get shipments of arms, ammunition, uniforms, and "advisers" landed by night on the American Eastern Seaboard under the very noses of the British. As I mentioned, it all worked rather too well. Freeing America, and bankrupting France. As a result of which, things French kind of fell apart, and they then had their *own* revolution, in spite of the best efforts of Louis XVI's director general of finances, Jacques Necker. Who cooked the books and almost convinced everybody the country wasn't really going down the economic tubes. But only almost.

Earlier on, in 1776, Necker had been running the Departement de l'Herault, in southern France, and had been approached by the young Swiss inventor of a new method for distilling. This was Aimé Argand, and a few years later, in 1780, he lit the world with a new kind of oil lamp. It used a circular hollow wick, and included the means to let a current of air keep the flame bright, and a glass chimney to keep the flame steady. The lamp put out the equivalent light of eight candles and was soon used to save lives near lighthouses, to illuminate the night shift at the Boulton and Watt factory in Birmingham, and (in February 1785) to stupefy audiences at London's Drury Lane Theater.

The whole idea of footlights had not long before been pioneered by Drury Lane's actor-manager David Garrick, whose amazing portrayals of Shakespeare introduced audiences to the modern, realist style of acting. And impressed a young Austrian painter called Angela Kauffman so much that she probably had an affair with Garrick and certainly painted his portrait. And that of almost anybody else of note in England. Kauffman was a recent and very beautiful arrival in London, from the hotbed of nouveau Neoclassicism in Rome. And she and her new style of painting took England by storm. Not quite storm enough, though. In 1773 when she applied for the commission to decorate Sir Christopher Wren's masterpiece, St. Paul's Cathedral, sorry, but she was foreign and papist and no thanks.

Wren himself wouldn't have cared. He was so High Anglican he was practically Catholic. He was also one of those Renaissance Men people say never really existed. Maybe not, but the breadth of his scholarship was nothing short of broad. Apart from inventing pens that wrote in duplicate, and odometers for carriages, he was expert in math, barometric studies, architecture, archeology, exploration, scientific illustration, and city planning. He also produced an algebraic work, written when he was young and relating to the Julian Period.

This period (about which I understand very little) was the basis for a great new calendar, dreamed up in 1583 by Joseph Justus Scaliger, French scholar and peripatetic Protestant exile, into which all historical events would fit because it provided one giant chronology for the whole of time. Point being that this would help historians agree on dates instead of having to argue about local references such as: "In the tenth year of the king's reign," or "ten days after the great inundations," or "the day of the new abbot's arrival."

Scaliger calculated his period by multipying the twenty-eight-year solar cycle (when, in the Julian calendar, the dates recur on the same days of one seven-day week) by the nineteen-year lunar cycle (when, according to the Julian calendar, the phases of the Moon recur on the same days of the week) by the fifteen-year "indiction" cycle (based on Diocletian's tax census period). I told you I didn't understand it. Multiplying these numbers gave a total of 7,980, and Scaliger reckoned the last time all three cycles had coincided (and therefore the last time his period began) had been January 1, 4713 B.C. This date was, therefore, Scaliger's "Day One" in spades. The date from which all dates could now be calculated in a uniform manner. And if *you* understood any of that, write me.

Scaliger had learned his stuff at the great Calvinist Geneva Academy, and, while he was there, never met Isaac Casaubon, with whom he was later to strike up an undying scholarly friendship (via more than a thousand letters), first on the subject of Greek and Latin manuscripts and then everything academic. While Casaubon was in Geneva, in 1556 he married (and had eighteen children by) Florence, daughter of one of Europe's greatest editors, Henri

Estienne, whose family had been in printing almost since the start, when his grandfather had opened a Paris print shop in 1502. Estienne's greatest claim to fame was that he discovered, translated, and, in 1554, printed the work of a sixth-century B.C. Greek poet named Anacreon, who wrote mainly erotic poetry and drinking songs.

Estienne's translation caught on all over Europe and by the eighteenth century Anacreon's poetry was so well known and enjoyed by well-heeled fun-lovers of London that in 1776 they formed a club named, after the poet, the "Anacreon Society." Aim: meet once every two weeks, get drunk, and sing songs. That's how glee clubs began. Anyway, one of the society members was a now-long-forgotten singer and composer (who went by the memorable name of John Smith), and when it was decided that the group should have a society signature tune, Smith whistled one up, entitled: "Anacreon in Heaven."

The tune was soon on the lips of everybody from tipsy clubmen who had survived a night out in London, to a nervous young American lawyer who had survived a night out in Baltimore. Well, September 13, 1814, *had* been quite a night, during which the British fired eighteen hundred shells at Fort McHenry. Watched from an offshore boat by the lawyer in question, who was so taken with events that he dashed off a commemorative song on the back of an envelope and set it to John Smith's tune: "Anacreon in Heaven." Today, it's better known in the version I was listening to while I watched the flag wave a few weeks ago in Washington, D.C.

Because Francis Scott Key gave Smith's tune a new name: "The Star Spangled Banner."

Inspiration is now flagging, so I'll stop.

10
THE SILK CIRCUIT

IN THE MID-SEVENTEENTH century one of the few places in northern Europe where the shopper could find high-quality silk was Spitalfields, in London. So in 1668, a Dutch draper called Anthoni Thonisoon went there to see the latest English designs. He was astonished to come across drawings of silk fibers magnified to a much greater extent than was possible with the draper's glass he normally used to examine cloth.

Fired by this amazing discovery, he returned to his home town of Delft, changed his name to van Leeuwenhoek, took up lens-grinding (the contemporary equivalent of computer software design), and began to mingle with the local scientific elite. On Christmas Day, 1676, his work burst upon an astonished English Royal Society in the form of a long letter containing illustrations of tiny objects Leeuwenhoek had seen through one of his five-hundred-power lenses.

What shook everybody was his statement that the objects appeared to be alive. With this first sight of rotifers and their waving cilia, protozoa cells rupturing, hair emerging from its roots, wriggling spermatozoa, and organisms thirty million of which Leeuwenhoek estimated would fit on a grain of sand, a new world opened to science.

For one German passing through Holland, the little critters were proof of the "Great Chain of Being" theory. This held that all life forms, from the simplest slime all the way up to humankind, had been designed by God in a series of successively more complex species that differed from their adjacent neighbors on the scale only by infinitely small gradations. The German in question, Gottfried Leibniz, had a vested interest in the microscopically small, having recently developed an infinitesimal calculus with which to work out the rates of acceleration of planetary bodies. Leibniz saw Leeuwenhoek's organisms during a visit to Delft in 1678 and asserted they proved that the differences between species might be so small "that it is impossible for the senses and imagination to fix the exact point where one begins or ends."

Leibniz's theory, based on the existence of such infinitesimally small, fundamental elements of existence (he called them "monads"), appeared to form the universal substrate the eighteenth century was looking for, in the wake of Jean-Jacques Rousseau's call for a return to the life of the noble savage, and the general disenchantment with the social effects of the Industrial Revolution, both of which had given rise to the search for a way to reunite man with nature. In the German city of Jena, the hotbed of this new Romantic view of life, Friedrich von Schelling's *naturphilosophie* brought together the recent scientific discoveries (of opposite poles, positive and negative electric charge, and acids and bases) into a unified theory of nature, which involved the dynamic resolution of mutually conflicting forces. Or some such.

It was in 1820, when attempting to apply this "conflict" view to electricity and magnetism, that a Danish wigmaker's apprentice, Hans Christian Oersted, forced more electricity into a wire than he thought it would take. The wire became incandescent, convincing Oersted that electricity and light must be related, so he extended his investigations and discovered that a current would affect a magnetized needle.

Twenty-one years later, this electromagnetic principle led Samuel Morse (among others) to develop the telegraph. In 1842, Morse provided Sam Colt (of revolver fame) with the means to trigger the explosion of one of Colt's new underwater mines, so as to demonstrate their effectiveness to President Tyler by blowing up a ship on the Potomac. Colt's other aim was to impress the Russians, who were also interested in mines. But because Colt was unwilling to explain exactly how his detonation process worked, the Russian contract went instead to a Swede called Alfred Nobel, whose mines needed no electric signals to trigger them. When a ship's hull hit Nobel's mine, the impact distorted a lead casing, breaking a glass tube inside the mine and releasing its sulphuric acid contents onto a mixture of potassium and sugar, thus causing a flame that ignited gunpowder.

During the Crimean War, the Russians sowed these new Nobel mines in the port of Sevastopol, forcing the Allied supply fleet to ride at anchor outside the harbor. This left the ships disastrously exposed to the great hurricane of November 14, 1854, during which the entire fleet sank, and with it the winter supplies for the army ashore. The winter deprivations that followed were so dreadful, Florence Nightingale's subsequent investigations brought down the British government and inspired Jean-Henri Dunant to found the Red Cross.

But it was the loss of the warship *Henri IV*, pride of the French Navy, that had the most long-range effect. The day after the disaster, French emperor Napoleon III called for the establishment of weather forecasting services throughout the country. By 1860 daily telegraphic weather reports were being published all over Europe. One of the leading figures in meteorology became a young American naval officer, Matthew Maury, who, over nine years, collated weather reports from all over the United States and amassed the equivalent of a million days' observations. From these he was able to prove that storms were all either circular or oblong.

By the 1930s the U.S. Weather Bureau had been gathering data for over seventy years and nobody had yet attempted to analyze the material. So a young physics teacher, John Mauchly, decided to attempt the task. The problem was: How long it would take to analyze the mass of data by conventional methods? Then Mauchly discovered that researchers studying cosmic rays were counting their particles using a vacuum valve. The valve could be made to turn on and off, in reaction to subatomic particle-strikes, up to one hundred thousand times a second. Mauchly realized that vacuum valves might automate the business of calculation by acting as data-storage devices.

Before he could do much about the idea, World War II broke out. Mauchly was conscripted and soon found another mathematical problem taking too long to solve. This was the calculation of artillery tables, used to instruct a gunner how to aim and fire his piece, under all conditions. Early in the war, at the U.S. Ballistic Research Laboratories in Aberdeen, Maryland, dozens of women mathematicians working around the clock could take up to thirty days to complete one artillery table for one gun (a single shell trajectory took 750 multiplications, and a typical table of calculations for one gun involved three thousand trajectories). By 1942 the laboratory was being asked to calculate new tables at the rate of six a week, so the situation was critical.

Mauchly put forward his vacuum-valve-counting idea and the Army accepted it. The process involved switching sets of ten vacuum tubes on or off, and using the total of each decade-set's on/off state to represent a number. Mauchly's machine was operative by 1946. Too late for the war effort, but not too late to calculate how to cause an atomic explosion. Called ENIAC (Electronic Numerical Integrator and Calculator), the machine was effectively the world's first electronic "computer," so named after the women mathematicians at Aberdeen.

ENIAC was fed its data by means of punch cards, whose adaptation by Herman Hollerith for use in the 1890 U.S. census had been suggested to him by his textile-industry brother-in-law, who knew about an automatic weaving system involving sprung hooks pressed against paper with holes in it. Where there was a hole, a hook would pass through and pick up a thread. Hollerith replaced the hooks with electrified wires, and the paper with cards. Where a wire passed through a hole representing a piece of census data, it would make electrical contact and cause a dial to move forward by one number. The system greatly sped up the census, counting 62,947,714 Americans in one-twentieth the time taken (for a much smaller population) by the previous census.

The weaving technique that Hollerith modified had been previously used to automate the production of cloth made of a material too expensive to make mistakes with: shot silk.

11

OUT OF GAS

I WAS TAKING a few days off in Swizerland recently, driving along the Geneva lakeside, when my rental car said it needed gas. And things being the way they are these days, a talking automobile didn't strike me as science fiction. Which was ironic, given the fact that at that very moment I was passing Villa Deodati, the place where another Brit holiday-maker invented the genre.

Back in 1816 Mary Shelley was staying at the Villa with her poet-lover Percy (they would marry the following year), smoking dope and having a generally unconventional time, together with their new pal Lord Byron and his young mistress (Mary's step-sister, Jane). One night the dinner-table conversation got round to the subject of how to revive corpses, whether or not artificial human beings could be made from separate parts, and the mind-boggling rumor that the well-respected Erasmus Darwin had apparently electrically "galvanized" some vermicelli and made it come alive. Well . . . who knew? Those science dingbats were starting to tamper with the fundamental forces of the universe and where would it all end, eh? So Mary, probably also influenced by the Humphrey Davy chemistry lecture she'd been reading (and Davy's remarks about scientists being able one day to discover the hidden secrets of nature), decided to write a cautionary tale about a young Swiss nerd, name of Victor, whose experimental mix of chemistry, physiology, and electricity goes horribly wrong and creates a monster Victor can't control. You'll have seen one or other film version of the story, I'm sure. I prefer Karloff's.

Mary got many of her technophobe views from her novelist and ex-preacher father William Godwin, the founder of socialism and political guru to such other left-wingers as Coleridge and Charles Lamb. Godwin (and most of his Romantic fans) reckoned the new factory lifestyle being introduced everywhere by the Industrial Revolution and its juggernaut machines would degrade the workers (only recently arrived in the rapidly expanding cities from their idyllic life in country villages) with diabolical and unnatural practices that included clocking-in for regular hours, regular wages, and shift-work. Godwin wrote yards of scribble on the subject of individuals being (mis)shaped in this way by their environment, and how the aim of industry should not be mass production and gigantic cities, but decentralization and egalitarian communities built on the human scale. And all this, long before anybody from Chicago ever said small was beautiful.

One of Godwin's most devoted groupies was a young Welshman called Robert Owen, who was to take Godwin's principles a good bit further. Owen was a spinning-mill superintendent in Manchester, so he'd seen the worst of what factory life could do. And in case I haven't made the point, despite the democracy of possessions that machines have given us in the modern world, it shouldn't be forgotten that the pleasures of consumerism were initially paid for by the most indescribably terrible living and working conditions in the early-nineteenth-century industrial cities.

In 1800 Owen became part-owner of a textile mill at a water-powered industrial site on the River Clyde, New Lanark, the largest single manufacturing enterprise in Scotland, where two-thirds of the machine operators were orphans. Extraordinary as it may sound to the modern reader, the use of pauper children in mills was considered real *pro bono* stuff, because without employment at the mill those little kids would have been starving, criminal, or worse. Which is why, when Owen and his partners took over New Lanark, by the standards of the time it was already a dangerously liberal place—the children's bed straw was changed once a month, they had two hours of school a day after work, their clothing was washed every two weeks and they slept only three to a bed, only seventy-five to a room.

Owen turned the place into a socialist utopia, with intervals for music and dance, an Institute for the Formation of Character, a company store, and a canteen. He also upped the working day to fourteen hours. By 1824 these libertarian ideas were catching on and Owen was heading a national labor movement, hob-nobbing with such admirers as Grand Duke Nicholas of Russia, reformer Jeremy Bentham, and the archbishop of Canterbury, as well as setting up a communistic settlement in New Harmony, Indiana. The fellow who'd sold him New Lanark in the first place (and whose daughter he married) was David Dale, a successful textile manufacturer who'd married into banking and one of whose very few failures had included a partnership in a Scottish cotton mill with one George Macintosh.

By 1777 Macintosh was making a scarlet dye called cudbear. Principal ingredients: ammonia (he collected urine from friends and factory workers) and lichen (in due course, when he'd scraped the Scottish Highlands bare, mostly imported from Scandinavia and Sardinia). Cudbear was a cheap alternative to the much more expensive color, made from cochineal beetles all the way from Mexico. The other thing about the cudbear dye was that alkali made it turn blue. In acid it went back to scarlet. Sounds vaguely familiar? When used by paper-dyers it was known as "litmus."

It was probably the cudbear link with ammonia that, in 1819, got George's son, Charles, involved with the Glasgow gasworks when he did a deal for all the coal tar they were throwing away. Coal tar was a by-product of the manufacture of coal gas, given off while cooking coal. In those carefree pre-ecology days, the tar was happily disposed of by the ton in quarries, rivers, and ponds. From this foul-smelling (and for the moment, virtually free) gunk, Charles Macintosh distilled out ammonia for the cudbear works (where production was, by now, far outpacing the well-meant but limited supplies of urine). He also recovered from the tar another chemical that would change the meaning of rainy days: naphtha. Charles found it would dissolve rubber and in 1822, after registering a patent, he spread the liquefied rubber between sheets of cotton and invented the raincoat (which to this day, in Britain, is still known—and wrongly-spelled—as a "mackintosh").

Not long after Macintosh was finding wet-weather ways to make money out of muck, a German chemist called von Hoffman received his doctorate in coal tar studies. With the nasty muck thus made academically respectable, more experimental things started to happen. In 1845 Hoffman was made director of the new Royal College of Chemistry in London and in 1856 one of his pupils, William Perkin, discovered the first artificial aniline dye (mauve) in coal tar. He did this while trying to make artificial quinine, but that's another story.

Meanwhile back in Germany a colleague of Hoffman's, named Runge, also experimented with coal tar and produced creosote, thus saving the American forests, because it preserved railroad ties so they didn't need replacing so often. The creosote prevented wood rot. And, as it was to turn out, other rots, as well. In 1857 in Carlisle, England, a form of creosote called carbolic acid was being mixed with sewage to prevent decomposition. Eight years later the professor of surgery in Edinburgh, Joseph Lister, heard about this trick, got hold of some carbolic from the nearby Glasgow University chemistry lab, and mixed it with paraffin. Thus armed, Lister then took the unthinkable step—he deliberately broke the skin over a non–life-threatening compound fracture and performed surgery. When the usually fatal festering appeared, Lister slapped his carbolic muck on it and, miracle of miracles, the patient survived!

A year later Lister was shooting a carbolic mist into his operating theaters and accomplishing new and daring procedures. The idea of antisepsis caught on so well he even did it to Queen Victoria (she had an abscess). In next to no time, surgeons (who like a good laugh) were leaping into operation with cries of "Let us spray!" (Previously, appealing to the Almighty had been about the only way to hope for surgical success, in a profession whose other standard joke had been: "Good operation, but the patient died.") Sprays were soon being used for local anesthesia, and on perfume bottles.

Then in 1883 a German engineer called Wilhelm Maybach teamed up with an ex-gunsmith friend, and together they used the spray in a way that would literally change the way the world worked. Maybach used the spray technique to turn gasoline into a fine mist, so that it would more readily ignite inside a piston cylinder and drive the cylinder up and down. The rest of the machine this gizmo was attached to was eventually named after a girl (the daughter of the company sales director) who was a favorite of Maybach's partner. The partner was Daimler and the girl was Mercedes.

Maybach's carburetor is, I suppose, ultimately responsible for my driving along the Geneva lakeside and running out of gas. Like this essay, now.

12
ORDINARY BUFFOONS

THE OTHER DAY yet again I came across that phrase coined by Roger Bacon in the thirteenth century to describe the process of technological advance (and overused by people like me ever since): "We stand on the shoulders of giants." Well, maybe. But I like to think ordinary buffoons have a part to play, too.

I have in mind the sad case of an archetypal English eccentric, an eighteenth-century admiral of the English fleet who rejoiced in the name of Sir Cloudesley Shovel. Shovel has two claims to a place on the great web of change. One is that he invented the "Shovel Wig," a full-bottomed creation about which it was remarked at the time that it looked like "nothing so much as a loaf of bread on the head." Shovel wigs were so expensive to buy and maintain that their rich owners were called "big-wigs."

Shovel's greater claim to fame is that he suffered death by drowning. But in a manner so spectacular as to kick off a chain of events that would end up with one of those essential products of technology without which the modern world wouldn't be the nice place it is.

Shovel drowned one nasty night in 1707 while bringing his fleet back to England from Gibraltar. In spite of having no idea where he was and being surrounded by thick fog, he pressed on, dead ahead, for home. Unfortunately he was a bit too close to home (aka the southwest coast of England), so he hit the rocks. Everything went to the bottom: fleet, two thousand mariners, and Shovel.

Now, too many ships were being lost like that, because they were lost, like that. But with the highly profitable American colonies waiting to be exploited, this was just the time for investors to be sinking their money in transatlantic transportation that wasn't sinking. So a very large prize was offered for anybody who could come up with safer ways of getting places. And back. In 1765 a clockmaker called John Harrison provided a timely solution.

The problem, navigation-wise, was the fact that on a

planet that spins (at the equator) sixty miles' worth (or one degree of arc) every four minutes, noon happens four minutes later for every sixty miles you travel west of your home port. And vice-versa. So knowing the exact time back at home will tell you how late (or early) the sun- or a star-rise is, where you are. Simple multiplication will then tell you how many miles one way or the other that means, position-wise.

Harrison observed, perspicaciously, that pendulum clocks wouldn't be much help, so he used a spring. At one end of his (amazing, new) steel clockspring he put a little brass slider, so that in the various weathers encountered along the way, the brass would expand and contract just enough more than steel did (in the ratio 3:2) so as to keep the expanding and contracting spring the same length, whatever the temperature. Harrison's chronometer lost only fifteen seconds over a trip to the Caribbean and back. Which meant you could return to within four miles of your home port. So, no more wrecks. Well, fewer.

Harrison's clockspring *was* amazing and new, thanks to another clockmaker, Benjamin Huntsman, who had come up with a mystery ingredient (he never revealed the secret) added to clay so as to make crucibles that would take the fantastically high temperatures needed to remelt the old kind of steel that had been too brittle for people to make good clocksprings with, unless you remelted it. Now, with Huntsman's crucible, they could.

Huntsman's crucible steel was also so tough that if you sharpened it, you could cut iron as easily as cheese. Which was the lifelong ambition of an obsessive ironmaker, John Wilkinson (he made a set of iron coffins—three for himself and the rest offered as gifts to friends; built an all-iron church; paid his workers in iron money; and slept with an iron ball in his hand so that when he dreamed of a good idea he'd twitch, the ball would fall and wake him up, he'd make a note of the idea, and go back to sleep). In 1774 he used Huntsman's steel on the blade of a new cylinder-boring machine that would cut metal so precisely Wilkinson was able to make the kind of piston cylinders James Watt needed, accurate to "the thickness of a old shilling." And then use the same cylinders to drive his own (first ever) steam-powered blast furnace. Hence the Industrial Revolution.

And then the one in France. Because the other thing Wilkinson's gizmo would do was bore out cannon barrels that were thinner, more precise, and interchangeable. These were smuggled to France (with which, at the time, England was at war) disguised as "iron piping," and they made possible the development of horse artillery, because the new barrels were also lightweight.

This last feature was to give added value to the work of a French military type called Gribeauval, inspector-general of artillery, for whom Wilkinson made his barrels (his other customers were the Turks and Americans, both of whom the English were also fighting at the time). From 1770 on, Gribeauval began the total reorganization of French artillery, reducing to four the many different calibers of gun and standardizing everything from ammunition to gun-carriage wheel size. Thanks to Wilkinson, he now had lightweight, mobile weapons that could be rushed from place to place in battle. This was an unheard-of way to behave. Cannon were supposed to take all day to position, and then all the next day to move. So Gribeauval's traveling guns changed the face of war. And (when the idea was taken up enthusiastically after 1792 by an ex-artillery officer and innovation freak named Napoleon) the face of Europe.

In 1810, once Napoleon had used his English cannon (and, incidentally, English uniforms and English shells) to good effect, and was, as a result, comfortably ensconced on the imperial throne, he decided to drag French industry kicking into the nineteenth century by motivational means, setting up an Institute for the Advancement of the Arts (confusingly, this meant Science and Technology). His grand plan was to make France militarily independent, so that the next time he fought, his materiel would be as French as was the word for it.

Initially, therefore, it may seem odd that one of the first prizes to be awarded by the new institute went to a guy called Nicholas Appert, who bottled champagne. But Appert, realizing that no army marches on its stomach quite like that of the French, sealed vegetables in some of his bottles, stuck the bottles in boiling water for several hours, and killed the germs he didn't know existed. Months later, in the Caribbean, the French navy opened his bottled veggies and declared them to be virtually fresh, the answer to scurvy, and (in terms of supply logistics) a quartermaster's dream.

A little later, in Paris, some passing Englishmen happened on Appert's bottling patent and bought the rights. And because one of them had a pal who owned a tin-making outfit they switched containers. Which is why we have canned food today.

However, during their patent-acquisition trip, what should these Brit investors also come across but an even more interesting improvement to French industry. It was an automatic paper-making machine—pulp, auto-scooped onto a shaking, traveling wire mesh, passed between felt-covered rollers that squeezed out the water and was then hung up to dry. A process hardly touched by human hand. So: no more need for all those paper-mill workers (who were away running around Europe with Napoleon's horse artillery, anyway).

For some strange reason the French hadn't picked up on the idea, so those English buyers went home with this patent, too. By 1840 England was turning out lengths of material that made possible the wallpaper we all love to tangle with today. And providing opportunities for artists-cum-social-reformers like William Morris to cover the paper with back-to-rustic designs, which now began to adorn the walls of every decorous Victorian home.

Made more decorous by another new home improvement: lavatories. Thanks to three cholera outbreaks (and a hundred thousand deaths), sewers and water mains and sanitation ware had started appearing everywhere. And because of that continuous-process papermaking technique from France, the socially upwardly mobile were now also able to add to their colorful and hygienic new lifestyle one final essential: the toilet roll.

All . . . thanks to Shovel. So never mind the "shoulders of giants." Let's hear it for buffoons.

13
BREAKFAST THOUGHTS

I WAS RINSING dishes at the kitchen sink the other morning when it occurred to me that what I was doing was (like everything, if you look long enough) a perfect example of the strange way things in the modern world are ultimately linked to each other. Things like the jet of high-pressure water I was using to wash the cornflakes out of my cereal bowl. And the cornflakes.

There was once a very early high-pressure water system set up outside eighteenth-century Paris, supplied by the River Seine and driven by a contraption of such humungous proportions that the local village name was changed from Marly to Marly-la-Machine. The Machine in question was a river-spanning line of water-mill–powered pumps built to feed the ornamental fountains at the Palace of Versailles located a few miles away. Water would shoot up into the air at great expense in order to amuse the king and his various mistresses. Now, extravagances like fountains and mistresses are fine when your economy is in great shape. Which, by the late eighteenth century, France's was not. Come to think of it, neither was the king.

But everybody's cash flow improved in 1797, when Joseph Montgolfier, a balloonist papermaker (job descriptions could be like that back then) showed members of the new Republican government a device he had invented that would deliver the water to Versailles (and, more democratically, also to canals, irrigation networks, and city water supplies) virtually free of cost or maintenance, on account of its having almost no moving parts.

Montgolfier's Hydraulic Ram, along with its various escape valves, used the current of water in a river to compress air, which then drove spurts of water up, down, or sideways, as often as 120 times a minute. By the year of Montgolfier's death there were some seven hundred rams at work, all over Europe. None of these were at post–French-Revolution Versailles, however. No king, no fountains.

With all that high-pressure water, rams could deliver a lot

of lifting power. In 1850 an English engineer called William Fairbairn used a variant on the ram to raise into place, at a rate of two inches a minute, a series of massive twelve-hundred-ton, rectangular-section iron tubes (through which trains would pass) for the Britannia suspension railroad bridge across the Menai Straits, in Wales.

Fairbairn also hired a fellow by the name of Richard Roberts, who had invented an automatic riveting-control machine. The machine used perforated cards that Roberts had seen controlling silk-weaving looms (discussed, in another essay, in a different context). The cards were designed to block or permit the passage of sprung wire hooks. Those hooks passing through the holes would then lift the particular threads relevant to that part of the pattern, so that the weaver's shuttle could pass underneath them. Roberts used his cards in a similar way to control the choice of size, number, and position of rivet-holes to be punched in a particular stretch of girder. The same perforated-card idea was picked up again later in the century by a man called Herman Hollerith, who ended up going into business with some people who later changed their company's name to IBM.

Meanwhile, Roberts's riveting technique worked so well that it caught the eye of a hot-shot engineer, Isambard Kingdom Brunel. He knew his new monster iron ship, the *Great Eastern* (in which he was planning to use the same tube-girder structures that had performed so successfully on the Menai Bridge), would need at least three million rivets for the hull alone.

As it turned out, riveting was about the only thing that went well for the ship. The engineers built her broadside-on to the Thames, only to find that the river was too narrow for a proper launch, so, at a cost of months and megabucks (well, megapounds), they were obliged to use rams and slides and winches and cradles, all manner of kit. Even then it took six attempts before she was finally in the water. On the *Great Eastern*'s maiden run to America, so many public catastrophes had already befallen the ship that only thirty-eight out of three hundred passenger berths were taken. Then there was a final day's delay, due to drunkenness among the crew. Things continued to go from bad to worse and by 1865, the biggest ship in the world ended up with nothing better to do than to lay the latest transatlantic submarine telegraph cable. And then to lose it, when the roll of the ship snapped the cable. And then to find it again the following year.

The cable was still in good working order when the engineers finally joined the broken ends together, because its sheathing had been made of the newest wonder-gunk: gutta percha, a kind of latex sap from Malaysian trees. But gutta percha did more serious things than allow President Buchanan to open the transatlantic telegraphic link with a few statesmanlike remarks to Queen Victoria. It also provided the first electric light insulation in homes, put chewing gum on the market, and revolutionized the game of golf.

Up to this time, golf balls had been stuffed with feathers. The problem was, most of these "featheries" lasted about as far as the third green. Then, in 1848, along came gutta percha and a solid, moldable ball, spherical enough to go where you hit it, and hard enough to stay in shape for longer than one game. The new balls were very grudgingly accepted at the jealously traditional Royal St. Andrew's (the world's oldest golf club, and they know it), where the new balls changed the shape of golf clubs in more senses than one. "Gutty balls" were so cheap that, by the 1850s, the game was attracting trainloads of holiday-making Scottish working men. To accommodate the crowd, St. Andrew's was obliged to split each of its (very wide) fairways longitudinally into two, each half to be played in opposite directions. Which is why there are now eighteen holes on a golf course instead of the original nine.

An early game at St. Andrew's must have been one of those rare occasions when the hot air you hear so much on golf courses was literally all about hot air. Hot air being the reason why there were suddenly so many factory-worker wannabe golf champs keen to improve their swing. At the time, the River Clyde at Glasgow was trying to become the world's greatest shipyard—and failing. The one thing any would-be industrial region needed was coal, but although there were thousands of tons of it buried beneath the Scottish Lowlands, the stuff was so low-grade that it would hardly make toast, let alone smelt iron.

Then, enter James B. Nielson, boss of the Glasgow gasworks. In 1827, he invented the gas-fired, hot-blast furnace, which could generate hearth temperatures high enough to make any fuel burn. Using the Nielson process, you could not only make iron using the local junk-quality coal, you could make three times more of it than you could using the expensive stuff. Within a few years came the Scottish Industrial Revolution, Clydeside shipbuilders, and the sootiest cities in the world. And, because of the new, iron-driven wealth, some of the best golfers anywhere.

Apart from opening up Scottish coal mines, Nielson's new technique was also hot news for the previously impoverished owners of the Pennsylvania anthracite coal deposits, similarly hard to burn, but now suddenly profitable.

In no time at all, Pittsburgh was becoming Steeltown U.S.A., and railroads were being built to bring iron ore down from the Great Lakes (and in the process changing the whole of business thanks to new railroad administration systems, such as cost accounting, monthly returns, divisional structures, and departmental management).

The burgeoning anthracite-fired iron and steel industry was soon littering the Pennsylvania landscape with large amounts of used coke. As it happened, Pennsylvania was where the great English chemist and American Revolution sympathizer Joseph Priestley had spent his declining refugee years. And it was he who had noted that coke is a great conductor of electricity. Drawing on Priestley's discovery, a resident of Pittsburgh, Edward Acheson, experimented with a mixture of coke and clay in an electric furnace. By 1885, as a result, Acheson was producing the second-hardest stuff in the world, a material he named carborundum.

Not surprisingly, Acheson soon went into the abrasive business. The bits of carborundum that he stuck on grinding wheels helped Acheson win the contract to manufacture the lights with which George Westinghouse would illuminate the great 1893 Colombian Exposition in Chicago. Those same grinding surfaces are still used, only now they are usually resin-bonded to their wheel by a process involving a solvent called furfural. Furfural is a chemical you get when you add sulphuric acid and water at high pressure to a mixture of throw-away plant by-products: oat husks, bagasse, and rice hulls.

And the cobs you have left over, once you've made cornflakes like the ones that I was so deftly washing off when I set out along this particular set of connections.

14

STONES AND BONES

I REMEMBER ONE day a few years ago, working in the great British Library Main Reading Room, and having a silent lament at its imminent demise when everything would move, as it did in 1998, to new premises, whose layout would not encourage octogenarian emeriti (like the one sitting next to me) to snore in that endearing way so familiar to users of the Room.

I suppose it was an Italian who did us all the favor, really. The Main Reading Room was the brainchild of Antonio Panizzi, after he became keeper of books at the British Museum in 1831, when the place was still embroiled in various scams associated with the lottery that was supposed to pay for setting up the museum in the first place. Although it must be said that whatever money might have ended up in somebody's pocket, what remained was enough to do the job, and the place was opened in 1759, thanks above all to the single-minded obstreperousness of Dr. Sir Hans Sloane, of hot chocolate fame (see elsewhere). When he died, Sloane had the greatest collection of "curiosities" anybody'd ever seen, and left it to the nation on condition there was a museum to house it, and twenty thousand pounds for his heirs. Hence the lottery, and in turn, the BM.

One of Sloane's buddies in all this was a surgeon called William Cheselden, famous for removing gallstones in fifty-four seconds (well, without anesthetic it hurt), as well as for being the queen's physician, for knowing Newton well, and for a massive 1733 book on bones, titled thrillingly: *Osteographia.* The tome included illustrations done with a *camera obscura*, and was the reason why a similar opus had been done with *no* pics a few years earlier, in Scotland. Point being, (a) the Scots author had been Cheselden's student in London and (b) all that stuff I said before about Cheselden being a big cheese and a friend of the great and the good.

Mind you, Alex Monro, the Edinburgh anatomist-author in question, was no shrinking violet himself. In 1726 he was one of a small group responsible for getting Edinburgh its first medical school. And for setting a new fashion in body-snatching from graves by students who wanted to ensure they got their money's worth at his dissection classes. Monro finally calmed public outrage by doing a deal with the local law, which involved some dubious arrangements for regular delivery of the fresh corpses of recently executed criminals.

One of the other hot-shots Monro had learned his stuff from, while studying down south in London, had been Francis Hauksbee, the man who invented the amazing "influence machine." This consisted of little more than a large glass globe on a spindle. If you cranked a handle (that turned the spindle and spun the globe) and then held a hand against the globe, you got an electrified hand. And the mystery influence also went down threads and attracted stuff like lint and feathers. Hauksbee had developed this spinning-globe trick as a result of experiments to find out what you could do with evacuated vessels, because science at the time was, literally, a lot of fuss about nothing. Everybody wanted to see what you could do with a vacuum, especially now that Hauksbee's boss, Robert Boyle, had come up with a pump that would make you a vacuum whenever you wanted one.

By this time Boyle had played around with air long enough to have formulated the law (about the behavior of a gas at constant temperature) known as Boyle's Law, or in French, *la loi Mariotte.* You'll notice the strange way the French have of saying "Boyle." That's because Edme Mariotte said he discovered the law at the same time as Boyle (or before, if you're French). Fact is, in 1679 Mariotte relied heavily on Boyle's work, but never mentioned him. Mariotte spent most of his life doing similarly similar things. Kind of blind spot, if you like. Ironic that one of the few original things he did was discover the blind spot. In the same vein, on one other occasion he "confirmed" the work of a fellow called Pierre Perrault, who'd recently measured the rainfall in the Paris Basin and concluded that the Seine (and rivers in general) got most of their water from rainfall.

In 1697 Perrault's brother Charles, who was a dashing type and, effectively, French minister of culture, handed the world to Disney on a plate with a collection of folk-stories called *Tales of Mother Goose* (which included "Sleeping Beauty," "Little Red Riding Hood," "Puss in Boots," "Cinderella," and others). He also got involved in near-fatal arguments (as the French will do about their language) over whether or not modern writers were better than the ancient Greeks and Romans (Perrault: "Yes," other extremely influential *gros legumes* like Boileau: "No"). Things went internationally ballistic when Perrault said Plato was boring. You could hear the row as far away as Ireland, where it engaged the passing attention of one of Eng. Lit.'s more acerbic wits.

Jonathan Swift's life is a perfect example of what happens when you ignore that old adage: "Be nice to people on the way up; you may need them on the way down." If it hadn't been for care-packages from the Berkeley family he'd probably have starved. And never have become pals with the Berkeley son George who went on to become a bishop and power in American education (the place in California is named after him), and who published, in 1704, *A New Theory of Vision.* In which he came up with the revolutionary idea that what you see is not what you get. That the brain interprets what the eye sees by associating the signals, received by the sense organs, with things already known in the brain.

This became known as "associationism" and eventually excited one of those people you love to hate. The medical genius of whom I speak could read at age two, and had read the Bible twice by age four. By age twenty he knew French, Italian, Hebrew, Arabic, Persian, Turkish, and five more. Get the point? So you won't be surprised to learn that by 1799 (age twenty-six) Thomas Young was already professor of natural philosophy at the Royal Institution, and lecturing only on acoustics, optics, gravitation, astronomy, tides, the nature of heat, electricity, climate, animal life, vegetation, cohesion and capillary attraction of liquid, the theory of sailing, hydrodynamics affecting reservoirs, canals, piers and harbors, techniques of measurement, common forms of air and water pumps, new ideas on energy. Had enough? Young had also published a new theory that light was probably a wave, and had done the famous experiment in which he sent light through two adjacent pinholes to produce the now-familiar interference patterns, and had announced that the retina responded to all colors in terms of the three primaries. Then, ho hum, he turned his attention to (and cracked the problem of) hieroglyphics. Well, wouldn't you? But it's comforting to know that in 1814 Young didn't face (and off-handedly surmount) an impossibly difficult task. In this case he had a relatively easy ride, thanks to a sample of hieroglyphics that the Brits had snitched from Napoleon when they'd driven him out of Egypt, in 1801. What made this object particularly useful for Young was that the hieroglyphic text was carved right next to its Greek equivalent. So he was halfway there (since, of *course*, he also knew Greek).

Funnily enough, when I got a bit bored with snoring scholars in the old BL, or the indecipherable hieroglyphs in the scientific monographs you have to wade through in my job, I used to wander out and enjoy looking at how much easier it was for Thomas Young. Because straight out of the old library door and second right was that object I mentioned.

The one Young used to crack the code. The Rosetta Stone.

15 IS THIS ESSAY NOTICEABLY DIFFERENT?

I WAS POUNDING the Azorean pavements in the predawn dark yesterday and listening to the radio in my ear when the incongruity struck me. Here I was, in mid-Atlantic and plugged in to the usual semiliterate stuff on the BBC World Service, because back in the early 1940s somebody got fussed about B-29 bombers and their temperamental onboard vacuum valves, vulnerable to the shakes and temperature extremes pretty much SOP on any bombing mission. Hence the postwar Bell Labs work of Bardeen, Brattain, and Shockley, who produced the solid-state, drop-it-on-the-floor-no-problem transistor made out of germanium, got the Nobel, and brought enlightenment to the early-morning ear.

Solid-state amplifiers turned out to be just what people working on masers wanted, so they could get away from having to excite ammonia gas molecules to give off microwaves and graduate to the big time with doped crystals that would make possible the laser, once the molecules of these crystals got excited enough to give off light. The coherent beam spread so little you could shine it at the Moon and see where it lit. One reason for the excitement, in more senses than one, was a dopant called neodymium. If you pointed just a little bit of illumination at it, at the right frequency, the molecule would go sympathetically crazy and shoot off a ton of laser light. So you got a lot more bang for your buck, as it were.

Ironically, the discovery of neodymium, in mid-nineteenth century, was associated with another form of blindingly bright light. It was while looking for materials that would glow incandescent that the Austrian chemist Auer von Welsbach (whom I've mentioned elsewhere) came across neodymium among the so-called "rare earths." Auer was to make his name in 1885 by impregnating a cotton gas-mantle with a mixture of these earths, thus making gaslight bright enough almost to beat Edison to the punch. Well, not quite. But he kept gas shares high for a decade or so longer than you might have expected after the first light bulb. Even today, his mantle's still going strong, in your portable camping gaslight. Anyway, neodymium wasn't the rare earth Auer was looking for at the time, so he used it instead to help make a cheap replacement for the flints that were being used in automatic-lighting gas lamps and named, after him, "Auer's metal." Like many an inventor, Auer was a bright spark who got his name in lights.

The reason Auer was such a dazzler was possibly that he'd learned his stuff in the lab of the man who put the "Bunsen" in burner. And Bunsen had got his gas, like everybody else, thanks to the chicanery surrounding the work of the penniless eighth earl of Dundonald, in Scotland. He was looking for a way to avoid bankruptcy when he roasted up some coal (he owned a tinpot mine and not much else), lit the fumes that came off, and made one of history's great discoveries without realizing it. Like an idiot he mentioned the fumes to William Murdoch, James Watt's sidekick, who promptly snitched the idea. Dundonald eventually died destitute in a Paris garret and Murdoch went down in history as the inventor of coal gas. (And who said science was honorable?)

From 1813 on, gaslight began to change life with everything, as it lit up factory night shifts, evenings out on the town, an increasing level of literacy, and classes for the artisan at the new Mechanics' Institutes. As well as making the plumber's candle famous. Weighing one-sixth pound and burning 120 grains an hour, this humble illuminator became the standard by which the bright life was now to be quantified (that is, it became the official measure of how much candlepower gaslight produced).

One of the ways of checking brightness was with a new

gizmo called a photometer. This made sure that a gas company's clients were getting their money's worth. Some photometers were like small, double telescopes, in which a prism brought two images (of plumber's candle and gas flame) side by side in the eyepiece. Then you moved a separate lens to magnify one image to the point where it appeared to be as bright as the other. The amount you had moved the magnifying lens to achieve this told you how much brighter the gas flame was.

Key word in this process: "appeared." Which is where some Germans entered the story, with a law of nature that I bet you've waited all your life to hear about—the "law of the just-noticeable difference." In terms of light, this new law came to matter most with instruments used to check on stellar magnitudes (where perhaps humankind's awareness of just-noticeable differences had begun, back when the ancients had classified by how much one star was brighter than another).

The modern law was first generally applied in October 1850, by a professor at the University of Leipzig named Gustav Theodor Fechner (he invented psychophysics, and to this day, October 20 is "Fechner Day" among the faithful). However, the idea was originally his teacher's, which is why it's Weber-Fechner's Law, and the point of it all was to measure by how much any sensation had to be increased before the increase was noticed. E.H. Weber tested it on the sense of touch by asking weightlifters at what point they noticed the extra kilos on the barbell, and Fechner did similar things to the other senses. Between them, the two men showed that the just-noticable difference in any stimulus was a constant, relating to the level of the basic stimulus.

The whole idea of something *being* noticed had originated from the work (alas, not noticed by many of his contemporaries) of a colleague of Weber's brother, Johann Herbart, who was the first to use the phrase "threshold of consciousness." Herbart was a pedagogue and his obsession was with how people learned anything. He'd come up with the concept of a mass of experience one gradually accumulated. He called this mass the "apperceptive mass" (what else?). Any new experience that came along was referred to this mass, and if you'd had the experience before, the event was subconsciously noted as a ho-hum affair, and nothing to bother your awareness with. But when something even *partly* new happened (I think I'm getting this right, but it *is* nineteenth-century German psychophysics), then bingo it had crossed the magic threshold and you became conscious of it. You see the connection with weightlifters, I hope.

It was Herbart's friendship with a Swiss ex-farmer turned teacher-to-the-world, Heinrich Pestalozzi, that got Herbart into making perception scientifically definable, because, I suppose, Pestalozzi didn't know (and didn't care) how to. By 1802 Pestalozzi had published *How Gertrude Teaches Her Children* and Herbart had written his *Pestalozzi's ABC of Observation.* Pestalozzi was by now famous for his school, where the kids learned from experience. No books, no formal classes. Development, not training. Showing the children a mountain before showing them the word for it.

By 1806 there was a Pestalozzi school in Philadelphia, run by one of his teachers, Joseph Neef, and in 1825 Neef was head-hunted to start a school in the new utopian commune at New Harmony, Indiana. This had been founded by the British libertarian Robert Owen, thanks to money from a well-heeled American businessman, William MacClure. MacClure was a geologist in his spare time, and had published the first proper geological map of the United States in 1809.

It was in the tristate area of Missouri, Oklahoma, and Kansas that MacClure's map identified the deposits from which, in 1952, zinc ores would provide the first major supply of germanium, the element used by Bardeen et al. in the transistor.

I bet you wonder where I dug *this* one up!

16
SHOWTIME

Even though I've spent most of my adult life working for the small screen (television, that is), I can't resist that magic moment in the cinema theater when the lights go down, the pictures come up in total-everything-scope, and I am enveloped in surround-sound. Even on those occasions when what's on the screen is kids' stuff, as it was recently when I was the adult-in-charge at a visit to a Disney movie.

I found myself thinking, as I watched, what a pity is it that Hollywood so thoroughly sanitizes the plot lines: These days you hardly ever get to see the real thing. I'm referring in particular to the tales by the Brothers Grimm, from which Disney took so much, but which were not quite the saccharine stories we've been fed by the animators. In the Grimms'originals, Cinderella's sisters had their eyes pecked out, Rapunzel got pregnant, Sleeping Beauty included a touch of necrophilia, and the Big Bad Wolf ate both Red and her granny. Even the Grimms' own editor toned it all down for the second edition.

Of course, the two brothers were dealing with pretty raw stuff, in the folktales they had collected from all over Europe as material for their stories. Around 1806, they had begun copying down the oral fables passed around by shepherds, wagoners, tinkers, gypsies, and peasants. Over a number of years the Grimms listened to what they thought of as the ancient voice of the Fatherland: legends laced with violence, cruelty, racism, contempt for foreigners, authoritarianism. Stuff that would warm the hearts of Nazis over a century later.

The Grimms were part of the great back-to-prehistory craze sweeping Romantic Europe in the early nineenth century. This folklore movement drew its inspiration from the new science of linguistics and the work of a Welsh judge in Calcutta. William Jones's job was to apply British law to Hindus in British India, so he studied the local culture. During the course of his investigations, he came across the amazing ancient Indian language of Sanskrit and promptly spread the word about its words. These were so similar to comparable words in Greek and Latin (and Germanic and Persian and Celtic and even Armenian and Albanian) that Jones reckoned Sanskrit had to be nearly everybody's ancient mother tongue.

Germans went particularly ape about Sanskrit because its existence suggested that the Teutonic races might have a cultural heritage that extended back every bit as far as that of the French (who, in the Napoleonic Wars, were clobbering Germany at the time).

The Indo-European mania kicked off by the discovery of Sanskrit affected even the most respectable types. One of these was a mathematician named Karl Gauss, at Göttingen University with the Grimms. Gauss's musings carried a lot of weight because of his reputation as something of a genius. He'd earned his position at Göttingen in the pluck of youth because in 1794, at age seventeen, he had developed a trick for calculating a planetary orbit even if all you had were three quick glimpses.

Which more or less describes the discovery of the first asteroid, Ceres. On New Year's Day, 1801, the Italian astronomer Giuseppe Piazzi spotted the new heavenly bod, but watched it through only nine degrees of its orbit before he fell sick. When Piazzi's health had improved, the weather hadn't. By the time the Palermo sky was finally clear enough for him to look again, there was Ceres: gone. Now, the discovery of a new planet had caused quite a Eurostir. And the loss of it, even more so. But all was not gloom and doom. Thanks to young Gauss's fancy math (officially, his "method of least squares"), bingo, a year later, Ceres turned up exactly where Gauss's figures said it would be. That feat made him an instant celeb, with astronomical job offers from all over. He chose Göttingen.

One of Gauss's fans was so impressed by the whole affair

that he couldn't bring himself to follow tradition and give his own name to a new element he'd just discovered. The chap in question was a stout, hypochondriac, Swedish gourmet and ladies' man who gave chemistry lessons to royalty, who, in return, gave him a title. Baron Jon Jacob Berzelius went everywhere, wrote travelogues, and knew everybody worth knowing. And the next time you're enjoying a refreshing draught of H_2O, you might want to recall that it was Berzelius who came up with modern chemical symbology.

Big B. was also the hottest guy in Europe on the blow-pipe, a device resembling a set of bellows, which he used to raise flame temperatures to fifteen hundred degrees Celsius. Under such ferocious heat, every substance becomes incandescent and glows in a characteristic way that reveals its constituent elements and compounds. In this way, Berzelius was able to analyze all kinds of mineral specimens for such hip-hop types as Goethe. He also inspected meteorites, and arcane materials like ancient Egyptian mortar and a Canadian trapper's gastric juice. This was why, in 1803, he was well able to examine a strange stone, identify the new element I mentioned earlier, and in honor of Gauss name it "cerium" instead of "Berzelium."

Now, one of the things I love about this web-of-change stuff is just what we've stumbled across. Here we are, with this dapper fat man and his mystery stone, in northern Scandinavian nowhere, about to make yet *another* notable contribution to the modern world thanks to the activity of somebody thousands of miles away who probably never heard of him. I'm talking about that great inventor and even greater self-publicist, Thomas Alva Edison, and his two dazzling contributions to our story: the creation of the incandescent light bulb (deliberate) and the effect it had (accidental).

In 1882, the opening of Edison's power station on Pearl Street, New York, must have made it clear to even the dimmest investors that they should get out of gaslight company stocks, which were now likely to go nowhere except out. But wait. Suddenly, another aristocrat-to-be appears on the scene, armed with a brilliant solution for the gas companies, thanks to Berzelius and his mystery stone.

Auer von Welsbach of Vienna discovered that if you impregnated a gauze wrapping (or "mantle") made of Sea Island cotton with certain minerals, one percent of which was good old cerium nitrate, and then stuck this gauze around a gas flame, the cerium glowed enough to increase by seven times the light ordinarily given off by the flame. That 1885 invention, known as the Welsbach gas-mantle, netted Welsbach an Austro-Hungarian baronetcy and the right to choose a (stupendously expectable) family motto: "More Light." The new mantle kept the gaslight business in business up until World War I (and you can still see it around some campfires to this day).

Some time around 1900, in a seedy Chicago rooming house, a fellow with the improbably Hollywood name of Lee De Forest was watching a Welsbach light and got the idea for a flame detector. He envisioned a gizmo that would pick up variations in the charge from an electrified gauze mantle, as the current in the mantle was affected by the presence of a flame. De Forest's problem was how to make the imperceptible variations in current big enough to perceive. So he co-opted a strange phenomenon that Edison had accidentally stumbled across when he noticed his hot light bulb filament giving off particles that dirtied the metal base plate of the bulb.

The phenomenon—which Edison named the "Edison Effect," then patented, filed and forgot—was eventually discovered to result from the steady stream of electrons that boils off the filament and rushes headlong for the base plate. De Forest placed a metal gauze between filament and base plate, and then used the powerful flow of electrons to magnify any tiny electrical charges that might be coming through the gauze. He called this all-purpose electrical signal booster an "audion." In 1914, the first New York–San Francisco phone line opened, audion-amplified all the way.

Now comes a real bit of irony. One of the many clever things Baron Berzelius had done, a hundred years earlier with his blowpipe, was to isolate another element, which he called "selenium" and which turned out to be useful in electric circuitry. One day in 1873, an operator at the transatlantic telegraph cable terminal on the Irish island of May noticed that the current in his selenium resistors varied strangely: high in sunlight, low in the dark. Turned out that the selenium was giving off electricity in response to light.

From 1900, a Professor Tykociner at the University of Illinois worked on a selenium-based voice recorder. He shone a light (whose intensity varied according to the vibrations of a microphone membrane) through film stock, to make negatives of varying exposures. When the film was printed it would let varying amounts of light through, to hit a selenium cell, which in response gave off varying current that then recreated the original sound on a loudspeaker. This was great, except for the fact that the signal was too weak to hear. Then voilà, in 1923, came the audion, in the form of the new Movietone system from Western Electric. And suddenly movies became talkies.

Which is why, thanks in the first place to the Brothers Grimm, I heard every word of one of their tales at the Disney movie the other night. That's all, folks!

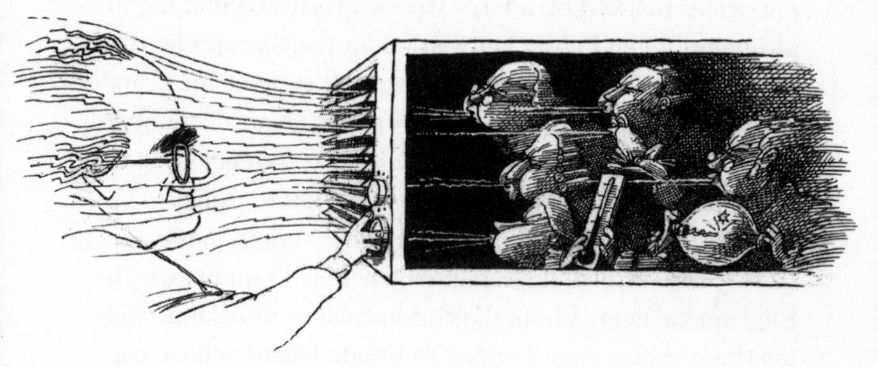

17

COOL STUFF

I WAS FIDDLING with the controls of an air-conditioning unit in a hot American hotel room recently and wondering how Celsians (that is, everybody else) manage with the Fahrenheit scale. Mind you, the man himself didn't have it much easier. By the time he was making his mercury thermometers in 1714 or so, there was a high degree of confusion about what was hot and what was not. Some contemporary thermometers had as many as twelve different scales marked on a large wooden board behind the instrument! When Fahrenheit settled for male-armpit-heat as 98.4 and melting ice as 32, life got a bit simpler for everybody. Fahrenheit's system was adopted universally enough for life also to become more precise. For science, that is, now that almost everybody could agree not to differ.

Fifty or so years later, thermometers were to make a bit of a splash in oceanography. When Ben Franklin was in England, he heard about the curious matter of Atlantic sailing times. Ships from London to Rhode Island, whose captains had to follow the entire southern coastline of England before striking out across the ocean, were taking two weeks less to get to America than ships taking the much shorter route from Falmouth (in the far west of England) to New York. Franklin's seafaring cousin Timothy Folger told him why. The Rhode Island packet crews were acquainted with whalers who knew about the 3-mph eastbound Gulf Stream, that would sometimes drive a westbound ship backward faster than her sails carried her forward. So, knowing this and avoiding it, the Rhode Island boats got across the ocean at a faster lick. In 1775, on his way home, Franklin lowered his thermometer into the water, found the Gulf Stream to be six degrees warmer than the surrounding sea, and produced the first chart of the current.

It was during his London stay that Franklin formed a lifelong attachment with a young woman named Mary Stevenson, when he and his son William lodged in her widowed mother's house. It's thought likely that he had wedding bells in mind for Mary and William. Well, it never worked out, as Mary ended up marrying a guy called Hewson. But she and Franklin corresponded regularly after he had returned to the United States, and in 1786 Mary moved to Pennsylvania to be near him, and to nurse him through the last days of his life.

This was nearly twenty years after her husband's own death, from an infection caught during a dissection. William Hewson had been a successful surgeon-anatomist, and spent over two years assisting one of the great surgeons of the day, John Hunter, who had a school of anatomy in London with his older brother William, the other great surgeon of the day. John Hunter was one of those guys you'd love to meet: a self-educated man who, among other things, wrote a natural history of teeth, discovered hearing organs in fish and the lymphatic vessels in birds, founded surgical pathology, was obsessed by hedgehogs, studied Portuguese geology, and married a woman who wrote librettos for Haydn (one day I'll follow *that* connection up).

John Hunter taught everybody who was anybody, including a young country surgeon named Edward Jenner, who lived an extraordinarily dull life except for the moment in 1798 when he used liquid from cowpox pustules to vaccinate people against smallpox and became humungously rich as a result (one of those "grateful nation" deals). Smallpox had been a killer for decades, and, before Jenner, entire hospitals were dedicated to not curing sufferers. Back in 1747, one of these, the Middlesex Smallpox Hospital, had a new high-tech ventilating system installed. Based on the design of an organ bellows, it had been invented by one of the hospital's governors, a public-health-conscious London vicar called Stephen Hales, whose other claim to fame was that he spent a lifetime studying why sap rose.

His electrifyingly titled work *Vegetable Staticks* shocked an Italian experimenter, Bologna University professor of obstetrics Luigi Galvani, into searching for the motive power that drove the human equivalent of sap around the body. He (and many others) thought there was some kind of mystery electric fluid running from the brain to the muscles, and that made them work. At one point in 1786 Galvani (what follows is not for the squeamish) hung a prepared dead frog, by a brass hook embedded in its spinal cord, onto an iron railing in his garden to see if "electrical" things would happen to the muscles during a thunderstorm. They didn't. At one point, however, in clear weather, when he pressed the hook against the railing, the frog's legs twitched! And the twitch also happened, to a greater or lesser extent, when he used hooks of different metals. Galvani immediately dashed off a terminally boring Latin tome announcing the discovery of galvanism, aka "animal electricity." Ten years later, in Padua, Alessandro Volta stacked alternate discs of silver and zinc (to reproduce the differing metals of Galvani's hook-and-railing arrangement), sandwiched the discs between discs of wet paper (to reproduce the moist frog tissue), attached wires to each end, and produced a spark. This proved that Galvani's current had been produced not by the frog but by metal-frog-metal juxtaposition. Volta's pile of discs (he called it a "pile") was the world's first battery. So let's hear it for frogs.

Curiously enough, Volta's other great excitement, besides frogs, was fogs. Well, miasmas. Well, bad smells of any kind. Most of which Volta found in marshland and turned out to be methane. If you've been around cows you'll know what I mean. Anyway, at one point in his smelly miasmic investigations Volta invented a kind of electric glass bomb. It was a corked glass vessel inside which two wires almost touched.

The other ends of the wires were attached to an electrophore (this consisted of a turpentine and beeswax cake, which, when rubbed with a catskin, would generate static electricity). First Volta would fill the glass bomb with a suspect gas (like methane). Electrifying the wires would cause a spark, and the effect would prove that the particular gas collected was explosive (big deal). It would also one day inspire the invention of the spark plug (really big deal). But why would a genius like Volta do a daft thing like that? Because everybody thought this kind of work would change the world of medicine. Miasmas (bad smells, dank mists) were reckoned to be the source of diseases such as malaria (which got its name from *mala aria*, the contemporary Italian for "bad air").

The other thing besides miasma that most likely caused malaria was heat. This was the view of one John Gorrie, a doctor living in Apalachicola, Florida, in 1850. He'd noticed perspicaciously that people living in cold climates never got malaria, whereas in the steamy swamps where he practiced, they frequently did. Since it was to be another thirty years before the word "mosquito" entered the discussion, Gorrie, like Volta, went off in the wrong direction. He built a small steam engine to drive a piston in a cylinder immersed in brine. The piston first compressed the air and then on the second stroke when the air expanded it drew heat from the brine. With each cycle the temperature of the brine dropped further. After a certain point the air in the cylinder was released into Gorrie's hospital ward to cool and cure his malaria patients.

Of course it didn't cure them. But, as the grandfather of all air-conditioners, the antimalaria machine ended up making my life bearable in that hot American hotel room.

OK, that's it. Off to chill out.

18 REVOLUTIONARY MATTERS

I WAS ENJOYING a recent partial solar eclipse in London and thinking about how, after Copernicus came out with *De Revolutionibus*, in which he made the shocking assertion that the Earth moved in orbit just like the other planets, for his contemporaries it really was, as they often said, "the world turned upside down."

Because if you put the Sun at the center of everything instead of the Earth, you rocked the entire boat and questioned everything: the old "fixed" order of things that could never change (and the church that said so); Man as the center of the universe (and the church that said so); the heavens that were beyond investigation (and the church that said so). No wonder Andreas Osiander (Lutheran mathematician and religious fanatic) tried to persuade Copernicus to do a smoke-and-mirror preface saying it was all just astronomer's mathematical fiction. Otherwise, said Osiander, there was a good chance Copernicus would be in deep and potentially fatal doo-doo with Rome. But since Copernicus was dying anyway, what did he care. In the event, while Copernicus's editor, Rheticus, was out of town (Nuremberg, where the work was being printed), Osiander, temporary replacement editor, slipped in his own preface, with the "fiction" message. By the time the thing blew up (Rheticus went ballistic) it was too late. And *De Revolutionibus* was off the hook enough to avoid the censor. More or less.

Osiander was a priest who dabbled in astrology and math, and at one point he corresponded with an Italian, Girolamo Cardano, who shared his interests. Cardano was quite a guy. From being in trouble with the Inquisition, he then became a pal of the pope. He also wrote over two hundred works on everything from music to philosophy to algebra to gambling. It was his luck with the dice (possibly because he came up with the first law of probability) that enabled him to pay his way through college. By 1540 he was already making a name as a math popularizer and algebra genius, dedicating his great work on the latter subject (*Ars Magna*) to his editor, Osiander.

Cardano was also the second-best doctor in Europe, and made one of the first detailed studies of asthma (which took him off for a strange adventure in Scotland, but more of that in another essay). We know that at some point Cardano met the *best* doctor in Europe, because he cast his horoscope. As a result of which it's said that Andreas Vesalius was born in Brussels at 05:45 on December 31, 1514. Vesalius advanced the state of medicine from potions of boiled puppy, lily leaves, and minced earthworms to modern anatomy, with his great new show-and-tell book titled *On the Structure of the Body* (1543) that literally took humans apart: brain, blood vessels, nerves, bones, muscles. The work was made easier thanks to a helpful judge in Padua, Italy, where Vesalius was Prof., who provided the author with fresh corpses of executed criminals more or less to order. The book was a boffo success, triggering an epidemic of grave-robbing by imitative medical-student wannabes.

Part of the reason why this new "what you see is what you get" approach made such an impact was its artwork, done by the studio of a new rising star in the elite Venetian world of portrait painting—Titian. And what Titian was all about (as any art historian will tell you) was the greatest skin-tone work anybody had ever seen. He even went so far as to make Madonnas and Venuses look like real, flesh-and-blood women (the kind that appeared in Vesalius's book).

Titian's stuff was so lifelike that, for instance, people doffed their hats when they passed his 1545 picture of Pope Paul III. So in no time the Venetian painter was turning down commissions right and left from queens, cardinals, dukes, princes, and other such would-be subjects. But there was one particular offer he just couldn't refuse. This was to paint the Holy Roman Emperor, Charles V, who was in Augsburg at the time, having just won a battle with the Lutheran princes, and who wanted a little commemorative number done. Titian sat him on a horse in full formal parade armor (the suit, made by his metal-tailor in Augsburg, that he'd worn to watch the battle). And you might conclude, on viewing the finished work in Madrid's Prado Museum, that the emperor was screwed. Well, that thing on his armor *might* be a wing-nut, but either way, it was how they fixed the suit onto you.

Augsburg, being a metalworking center, was the home of the screw. And one of the local goldsmiths was a fellow called Max Schwab, who sometime after 1550 got an invite to send one of his fancy new screw presses to the Louvre, in Paris, to make some not-so-fancy new coins for Henry II. These were to have a reduced precious-metal content, the idea being that devaluation would help to shore up Henry's tottering finances and also leave enough for his wife Catherine de' Medici to spend on her diplomatic mega-buffets. At one of which she might well have introduced a new, fun drug she'd been sent by the French ambassador to Lisbon.

If I tell you his name was Jean Nicot, you'll be there ahead of me. Tobacco went through European high society (and then the rest) like an addiction waiting to be taxed. Soon enough (when James I of England upped the tobacco customs duty by over thirty times) it became the first of many examples of a harmful but fiscally remunerative commodity. So, before you knew it, mercantilism was the new buzzword. Tobacco sales started pouring cash into national coffers, and the new European craze became: setting up colonies to make even more green stuff from the green stuff.

Unfortunately, this left seventeenth-century France potentially out of the get-rich mercantile game, since all they had for a navy was a few dozen rotting hulks. Until a new controller-general of finance, Jean Baptiste Colbert, took over, and turned the French economy around with everything from tax incentives for explorers (which is when French Africa started) to a new domestic transportation network. Including the Canal du Midi (still there for holiday boating), for which he engaged the services of (among others) a military engineer genius called Sebastian de Presle, Seigneur de Vauban, who did one of the aqueducts.

Vauban was quite a guy. He predicted what the population of Canada would be in A.D. 2000 (fifty-one million), and wrote on bee-keeping, silk manufacture, pig-farming, and taxation. He built fortresses all round the French frontier and invented the socket bayonet. He also came up with a totally new kind of siege tactic. We call it trench warfare: Dig a trench, fill it with firepower so as to cover the men digging the next trench, nearer the enemy walls. Go on doing that until you're right under the walls. Then undermine them with explosives and blow them up. Unless you've undermined enemy morale so much they give up before you have to.

Which is just what happened on October 17, 1781, when General Cornwallis and his troops surrendered the fort of Yorktown and marched out, signaling the victory of the Americans over the British. And you know what the band played as they left?

A little march entitled "The World Turned Upside-Down."

19

DON'T FORGET THIS ONE

I GOT ONE of those junk-mail fliers through the post the other day, urging me to sign up for a study-by-mail course on the subject of memory improvement.

"If only I knew where, or what, my memory was," I

thought, "I'd do anything to improve it!" Still, we know more than we used to about such neurophysiological matters, thanks in part to the kind of junk mail that was sent out in the 1850s by Isaac Pitman and his business partners in their efforts to promote a totali nu wei uv speling Inglish. Alas, their efforts came to naught (or nought), and they switched instead to selling correspondence courses for a phonetically based writing technique now known as "shorthand."

Pitman's original reason for attempting to turn English into WYSIWYG was that it isn't. (Try pronouncing "Featherstonehaugh," if your mother tongue isn't *English* English. Go ahead and try, even if you speak English from parts American, Australian, New Zealand, Canadian, or South African. Give up? It's "Fanshaw.") Pitman believed that world peace would be more rapidly achieved if, by making words such as "Featherstonehaugh" simpler to read and pronounce, all those foreign johnnies could be more easily exposed to the civilizing influence of *English* English. Harrumph. The idea took root, though on a much grander scale than the single-minded Pitman might have hoped for. In 1897 it flowered as the International Phonetic Alphabet. Which made every language easier to read and pronounce.

Top gun in phonetics was Henry Sweet, after whom George Bernard Shaw modeled Professor Higgins in *Pygmalion* (aka *My Fair Lady*). As it happens, in the play, Higgins notes down Eliza's speech patterns using another set of symbols, called "visible speech," which had been developed, long before, by Alexander Graham Bell's father, an elocution teacher who had been a founding member of the British Phonetics Council. In the 1870s Bell, Jr., was busy visualizing sound, too, for the deaf students he was teaching in Boston.

It was at this juncture that he came across a thing called a phonaudiograph, developed by the otherwise-almost-forgotten noodler Leon Scott Martinville. The device was fairly primitive: A membrane vibrated in reaction to speech, and a bristle attached to the other side of the membrane traced wiggly marks on a moving piece of smoked glass. With the phonaudiograph Bell was able to show his deaf pupils the correct "shape" of the sound they were trying to make, so that they could then compare their own attempts to imitate it.

The whole wiggly-line phenomenon probably had its origin in an invention years earlier by a French physiologist by the name of Etienne Marey, who fitted a membrane on a tiny drum (a "tambour") and placed this wherever he wanted vital rhythms to be turned into graphs. When pressure of any kind depressed the membrane, the air in the tambour would be forced along a tube to push against a membrane fitted to another tambour at the far end of the tube. A stylus mounted on this second membrane would move in response and trace a line. With the tambour (still in general medical use as late as 1955), Marey could reduce virtually any kind of movement to lines. Marey called his squiggles "the language of life."

That the innovation should have come from France made sense because in the early 1800s, Parisian hospitals were far in advance of anybody else's. Even the English came to take notes. It was in Paris that ward rounds and charts and stethoscopic diagnosis and medical statistics first became common. All that, and the unquestioned authority of doctors (stemming from Napoleon and the fact that he conceived of winning wars with a million untrained conscripts, many of whom inevitably ended up in the Paris hospitals, obedient to discipline and too ignorant to question anything that was done to them).

Numbers also did the trick for medical technology, because there were so many war wounded that hospital staff could now easily gather really large-scale, statistically meaningful amounts of data on the efficacy of diagnosis and treatment. And so those wiggly lines began to appear at the foot of the hospital bed, showing the progress of the patient's temperature, respiration, pulse, and heart rate, or any other physical condition that could be reduced to lines and figures.

Which, late in the century, left only psychological disorders to systematize. This began with a hypnotically exciting new Viennese technique called "Mesmerism." Franz Mesmer's assistants would first examine patients in order to locate their "magnetic poles." Then Mesmer himself would appear (in feathered hat and long robes) to stroke relevant areas of the patient's body, transmitting a curative, mystery "influence" to the sufferer.

Despite the fact that such hardheads as Benjamin Franklin officially pronounced Mesmer a fake, the concept of "influence" persisted. The idea had already, after all, been around for three hundred years. Even Descartes had thought a "vital spirit" flowed down the nerves from the pineal gland. For this reason, two more Viennese medics, named Franz Gall and Johann Spurzheim, were, by 1820, successfully huckstering a brand-new science: phrenology. It was based on the theory that if the "influential liquid" postulated by Descartes originated in thirty-seven separate organs in the brain (each responsible for a moral, sexual, or intellectual trait), then your character might be assessed by feeling the bumps in the skull above the organs. If the organs were particularly well-developed, there would be a bump in the skull above them. If you had such a bump behind the left ear, for instance, you were a good lover. (Did you just check?)

In 1876 an Italian, Cesare Lombroso, director of a lunatic asylum, had studied thousands of heads (living and dead) and found further evidence to support Darwin's theory of human descent from apes. It was Lombroso who spread the word about criminals and the insane being "throwbacks," with sloping foreheads ("Neanderthal" was the new buzzword, thanks to the recent German discovery of ancient human bones in the Neander Valley).

Conservatives took Lombroso's remarks about "criminal characteristics" to mean that criminality was innate, and from this inferred that a "born" criminal could not be rehabilitated. On the other hand, liberal thinkers such as Jeremy Bentham saw in bump-reading the opportunity for self-improvement, and urged prison reform.

Perhaps the most startling outcome of Lombroso's work was the effect it had on a young man who was briefly his assistant at the asylum in 1872. The assistant's job required him to carry out post-mortem dissections, and, intrigued by talk about bumps of knowledge and low brows, he began to slice brains in his kitchen and peer at them with the microscope that his pathologist uncle-by-marriage happened to have.

Some time in 1873, perhaps stimulated by reading about the new chemistry of photography, he left a piece of brain to harden in a mix of potassium bichromate and osmium chloride, then dunked it in a solution of silver nitrate. (Curiously, he couldn't be persuaded afterward to explain how he had arrived at this process. All he would say was: "I discovered a method by using the method which I discovered.") But it doesn't really matter how he got the idea. The important thing is that when he cut very fine slices of the material, dried them, and illuminated them from behind, he found something that would change how we think about how we think. What Camillo Golgi saw was a golden-yellow background of tissue, within which were visible, in glorious and finely detailed black, the brain cells that today bear his name. From that single experiment, the whole of modern neurophysiology was to emerge.

So if somebody ever succeeds in discovering how to improve our memory, it'll probably be thanks to Golgi and the phrenology phreaks.

And that junk mail I mentioned at the beginning (. . . remember?).

20 TAKE TWO ACRONYMS

NOW AND AGAIN, doing the reading in preparation for an essay can involve working your way through the kind of research material that would give anybody a splitting headache.

Here's one particularly gobbledygook example of why: "The mathematical chance that more males than females shall be born in a year is shown by theorem to be less than 1 to 2; but to make the argument stronger, it is stated as 1 to 2 for a single year. The chance that the same thing will happen for 82 years in succession is, then, 1 to 2 with the exponent 82, or very slight; and if not only 82 years but 'ages and ages' and not only London but all the world, be included, the chance becomes an infinitely small quantity, at least less than any assignable Fraction."

Well, *you* try reading John Arbuthnot's statistical study of male-female births in London between 1629 and 1710, and see if it doesn't give you instant catatonia or worse. Fortunately, the good Dr. Arbuthnot (physician to Queen Anne and the man who first said that the study of math was good for the moral health of young men) had a lighter side to an otherwise harrumph personality. In his spare time, he wrote a number of satirical pamphlets (about current British political efforts to achieve peace in Europe) that would put the "John Bull" character into the British national consciousness. This effort was part of his creative output associated with membership of an avant garde London literary society known as the Scriblerius Club. Once every two weeks, adepts at verbal vitriol (such as Alexander Pope and Jonathan Swift) would meet to send their fellow-clubmen into paroxysms with a scatological send-up of some particularly devious politico or other. Then they'd publish under the pseudonym of Martin Scriblerius and see what they could get away with.

One regular Scriblerian enjoying the fortnightly fulminations was a hedonist heterosexual named John Gay, who, on January 28, 1728, made his name forever when he ousted "Italian" in favor of "lyric" by staging the first night of his

Beggar's Opera. Talk about boffo. A straight run of sixty-two sold-out performances. And the talk of the town. Which is probably why Voltaire's pals (including Pope and Swift) took him to see the show, when the famous Frenchman was on a secret visit to London, keeping a low profile after a bit of a dust-up back in Paris with a well-connected aristo who'd had Voltaire mugged for having the temerity to suggest that the two of them have a duel over some imagined slight. (Well, Voltaire *was* common.) A year later, when the dust had settled, the eminent philosopher went back home to spend the rest of his life being a thorn in the flesh of anybody in power and, not surprisingly, a permanent fugitive from one or other French national chief of police.

He did, however, manage a few happy and relatively trouble-free years closeted away with the lovely Marquise de Chatelet at her castle in deepest Champagne. During this idyllic sojourn the brilliant couple (she was knocking off something on Newton's math and so was he) had the kind of effect you might expect from a better intellectual mousetrap. A path to their portcullis was beaten by everybody who was anybody. Plus one nobody, name of Karl-Victor von Bonstetten, a young German-Swiss nobleman on the Grand Tour and looking for a bit of enlightenment.

That's not all he was looking for. Shortly thereafter (this was in 1774) he talked himself into what's often referred to as a "leg-over situation" with the unhappy wife of a fat, drunken, pox-ridden, pompous old womanizer living in Italy and calling himself King Charles III of England. Which he wasn't, although it must be said that he'd been a serious contender for the title a few years earlier. Back in 1745, Charles Edward Louis John Casimir Silvester Maria Stewart had gone down in history as the dashing young Bonnie Prince Charlie, Scottish Pretender to the English Throne.

He even got his ragtag Highlander army within a few miles of London, before being let down by the French, who'd promised to support his coup with massive reinforcements from across the Channel, and then chickened out at the last minute. Ironically, the manifesto justifying this Gallic intervention in *les affaires britanniques* had been written by none other than Voltaire.

Anyway, one of the people who spirited Charlie away one step ahead of the sheriff, after the final and disastrous battle of Culloden (where the Pretender had lost because his men used claymores and the Redcoats used artillery) was a woman called Flora MacDonald. This intrepid Highlander smuggled Charlie out of the danger zone (to the island of Skye and a boat for the Continent) dressed up as a woman. As it happens, his transvestite alias was Betty Burke. No relation. Then, after MacDonald's inevitable capture and imprisonment by the English, Flora made her own miraculous escape to the Cape Fear area of North Carolina.

Where the only work to be found for many of the Highlanders who also fled there at the same time was in the production of naval stores for the British. You roasted the long-leaf pine trees (with which Cape Fear is still well-stocked) and they gave up resin. This could then be boiled or distilled into various materials such as pitch, tar, and turpentine that would make anything you liked waterproof. This included ships' hulls, ropes, planking, and your insides. This last, given that doctors at the time favored the ingestion of turpentine for most respiratory or dermatological conditions. Be that as it may, come 1776, when that whole unfortunate revolutionary thing happened, and we Brits lost N.C. (and the rest), a new source of pitch, tar, and turpentine had urgently to be found, or Britain wasn't going to be ruling any waves any longer.

This was the reason, shortly thereafter, why an impecunious Scottish earl named Lord Archibald Cochrane, whose family had backed the wrong king and the wrong horse for several generations, was roasting coal just outside Edinburgh. By this time Cochrane's patrimony was reduced to little more than a few small coal mines, and his idea was that this coal-cooking activity would solve the Royal Navy's pitch, tar, and turpentine deficit (and therefore that of Cochrane's bank balance) by producing lots of black stuff from which he'd make lots of green stuff. The very same day the noble lord turned up in London to offer his amazing new gunk-making process to the Royal Navy was, unfortunately, the very same day the Admiralty in London decided to copper-bottom all naval ships. So Cochrane was left with a useless load of sticky black muck, and would eventually end up dying in poverty in Paris.

Alas for the vagaries of history. Cochrane's muck turned out to be coal tar. And we all know what *that* turned out to be. Almost anything you (or a chemist) would care to name—artificial dyestuffs, phenol, antiseptics, creosote, pyridine, and I won't go on. Except to add that in 1890 one of the many compounds being discovered in coal tar was a couple of easy steps from phenol. A German chemist, Felix Hoffmann, working for Bayer, was to derive from phenol a substance called salicylic acid. No distance from there to acetylsalicylic acid. Now, in its natural state, salicylic acid comes from the plant meadowsweet (Latin: *spirea ulmaria*). So Hoffmann gave his new wonder drug an acronym—A (for acetyl), SPIR (for *spirea*), and IN (for a reason nobody knows).

The full acronym (line up the letters) solved my earlier problem with Dr. Arbuthnot. I'm going to lie down now.

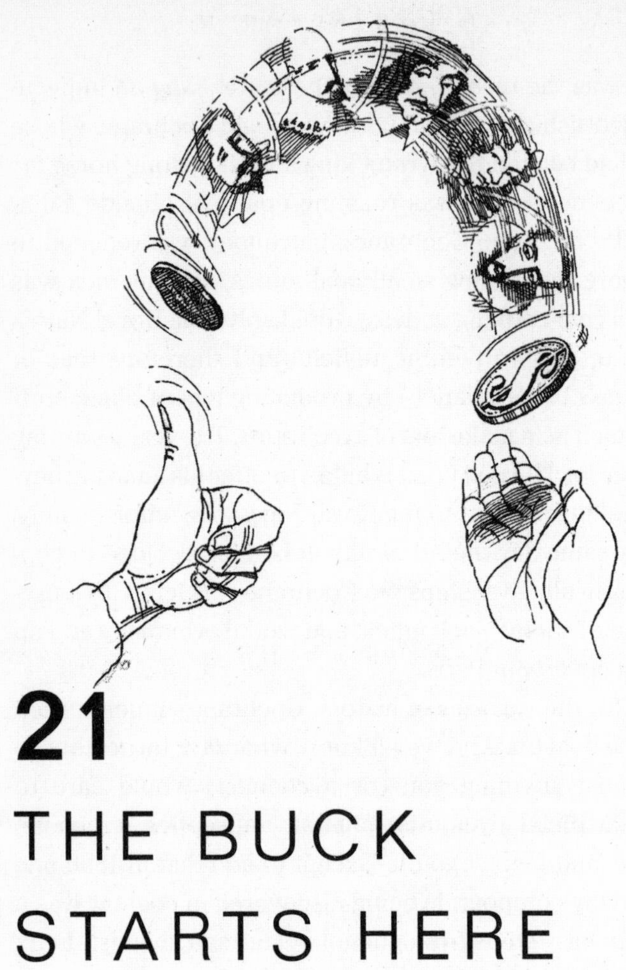

21
THE BUCK STARTS HERE

I WAS GOING Dutch the other day at lunch in New York, and handing over my share, when I remembered it was a sixteenth-century polymath from Holland who started all that decimal money stuff. Simon Stevin was his name. Unsung hero would be nearer the mark. His motto could have been that of any of the Scientific Revolution biggies who later eclipsed him: "There's always a rational explanation for what looks like magic." Stevin was the engineering genius who first popularized an alternative to the mind-wrenching medieval practice of calculating everything in fractions (for a flavor of the torture involved, try "3/144 × 2/322 − 1/85 = ?"). He turned such gibberish into the decimals that scientists (and even innumerates like me) could more easily work with. He even gave the treatise he wrote on the subject a user-friendly title: *The Tenth.*

In 1585 Stevin became tutor to Prince Maurice of Nassau, ruler of the northern bit of Holland at the time. Maurice was a bit of a military history freak and thought there was much to be learned from the way the Romans had fought. So he built an army that was, so to speak, all dressed up with nowhere to go. That is to say, he introduced new techniques (an instruction manual for firing muskets by numbers, the use of cartridges, and military drill) any one of which could have given him victory at any major battle he ever found himself involved in. But he never really got anything much more than a large skirmish. This left all the glory to a Scandinavian, King Gustavus Adolphus, who a few years later gave his soldiers even more winning ways. One of his key improvements was to put the musketeers in three rows, so that while the front row was firing, the second and third rows were reloading, ready to step forward and pull triggers. So there was a constant stream of hot lead heading toward the opposition. As a result of this trick Gustavus won every battle he fought (even the last, at which he was killed) and made Sweden a world power for all of fifteen minutes.

The subsequent Swedish ruler, Gustavus's daughter King Christina (that's not a typo—only the wife of a Swedish monarch was called "queen"), cut her hair short, wore men's clothing, turned Catholic, and abdicated in 1654 after only ten years in the post. Whereupon she hightailed it off to Italy and (it is suspected) a long-term affair with one of the cardinals in a troubleshooting team of elite Curia bureaucrats known as the Flying Squad. While some Swedes may disagree, there are many who believe we have much to thank Christina for. Among other things: Rome's first opera house, and the successful careers of Bernini, Scarlatti, and Corelli (all of whom she protected from various forms of Roman back-stabbing). Less admirable, perhaps, was what she did for René Descartes. While she was still queen (sorry: king), she invited that eminent French philosopher to come and be thinker-in-residence, and then obliged him to give her philosophy lessons at five in the morning. In Stockholm. In January. Surprise surprise, he caught pneumonia and died.

Fortunately, however, not before he'd produced (among others) the "Discourse on Method" that taught us all to think straight, as well as a fundamentally new view of the cosmos, and a major piece on how the human body functioned like a machine. This last included a bit on how the brain worked through a system of tubes and valves controlling the distribution of a fluid "animal spirit," that made the different parts of the body move. This moved a certain Tom Willis, rich and successful physician at Oxford University, to spend years preparing a major work on matters cerebral, titled *The Anatomy of the Brain.* Definitive for the next 150 years, it contained the first reference to the autonomic responsibilities of the cerebellum. Moreover, if you're a neurologist, you also have Willis to thank for your job description. Willis's book was an international bestseller because it was the first to feature copious illustrations so detailed and accurate even the *New England Journal of Medicine* would have accepted them.

The drawings were done by England's greatest draughtsman, architect, linguist, mathematician, weather forecaster, astronomer, and general big-head (well, at the age of only thirty-six you'd have to have a lot of chutzpah to apply for, and get, the contract to build St Paul's Cathedral after the Great Fire of London). Christopher Wren was also a canny businessman (you're not surprised), being one of the first to get into the new stocks-and-shares game and becoming a director of that license-to-print-money known as the Hudson's Bay Company. Considering how much profit this organization made (and still does) for its backers, it seems a pity the bay's eponymous discoverer did all the hard work for so little reward.

Like many early European navigators, Henry Hudson spent a lot of his life going nowhere. In particular, in 1609, when he was commissioned by the Dutch East India Company to find an Arctic Northwest Passage over the top of Greenland and America, so the Dutch could get to the spices, porcelain, and tea out in the Far East without being hassled by the Spanish and Portuguese who had the southern routes sewn up. Well, after sailing up and down the Greenland coastline for months, and repeatedly bumping into Spitzbergen or the pack-ice, Henry gave up and went back to Antwerp to give a piece of his mind to the so-called cartographer who'd put him on the road to nowhere. This unfortunate was theologian-turned-mapmaker Pieter Platvoet, who'd learned all he knew (Hudson: "Not enough!") from the truly great cartographer Mercator, whose fame spread rapidly when he was printed by Christopher Plantin, Europe's richest publisher with a profitable sideline in French underwear.

Plantin made a fortune when the Council of Trent decided to standardize worship and ordered more than forty thousand identical liturgical texts for Philip II of Spain. Or would have, if Philip could have paid his bills on time. Philip's little financial problem was his father, who'd got the job of Holy Roman Emperor by greasing the right palms with money he'd borrowed from (and left Philip to pay back to) a German banker named Anton Fugger, the Rothschild of the day.

By this time the Fugger family had been in the money game for over a hundred years, and pretty much every crowned head in Europe was in hock to them. The problem being that all these kings and princes used mercenary armies and never had the ready cash to pay them off. So the Fuggers would helpfully provide the wherewithal, in return for property, or tax breaks, or concessions. One such recoupment package included a mining franchise in the mountains of Bohemia. Where there was a hole in the ground that produced so much silver it became the official source of coinage for the entire Holy Roman Empire. The mine was in a valley named Joachimsthal and so the coins it turned out were given the same name: "Joachimthalers."

Over time this became shortened to "thalers." And over *more* time, the American pronunciation of the word became the name for the currency I was handing over at the end of that meal at the start of this column.

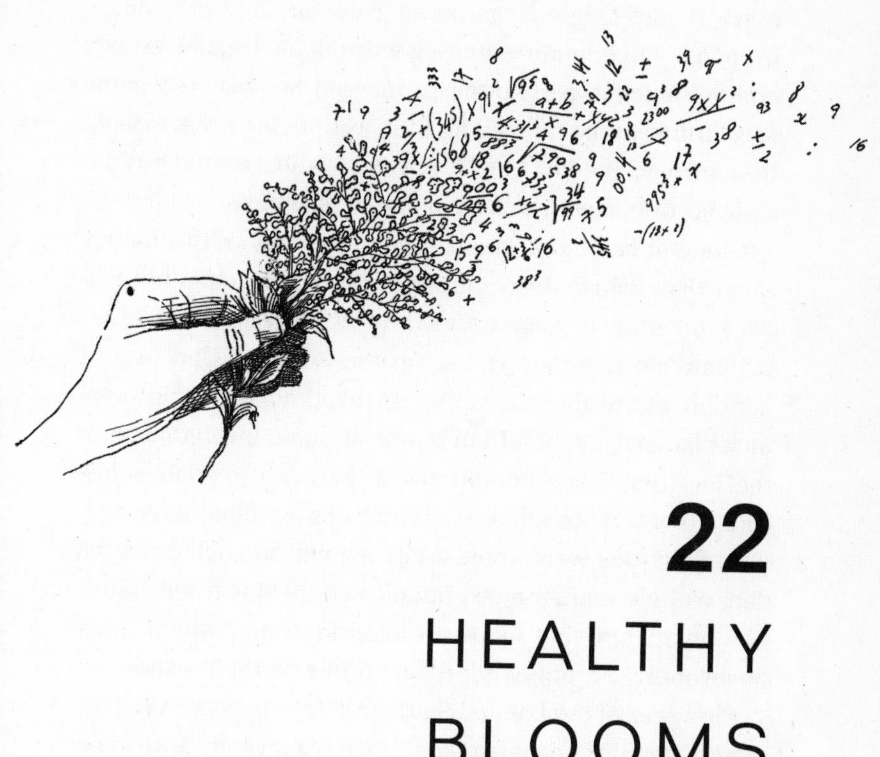

22 HEALTHY BLOOMS

I'LL RISK A bet. You (like me) didn't know that the common lilac flowers when the sum of the squares of the mean daily temperatures (Celsius) since the end of the previous frost add up to 4,264. This piece of mind-boggling botanical trivia sprang fully formed from the brain of a Belgian astronomer and mathematician named Adolphe Quételet, whose obsession with numbers led him in 1835 also to invent a concept that I'll bet you *have* heard of: the average joe. Quételet gathered data on this individual's propensity to commit crimes, get drunk, marry, die, be tall, commit suicide, and so on. In the end he found so many regularities in the figures, he said, as to believe there could be such a science as "social physics," which would put the analysis of behavior onto a mathematical basis.

Quételet had taken some earlier thoughts on this matter to an 1833 British Association for the Advancement of Science meeting in Cambridge, where he persuaded other like-minded noodlers to set up the Statistical Society of London and further the cause. At this time, British interest in social analysis of all kinds was at panic level, thanks to the fact that living conditions in the overcrowded industrial cities had brought the laboring classes close to revolution. Statistics were soon avidly sought on such essential data as how many ragged families could sing a jolly song, how many starving mothers could knit, and which filthy hovel sported morally improving prints on their walls.

First prez of the London Statistical Society was one of the nerds Quételet had met in Cambridge, a man with more ideas than time to do much about them. These included a speaking tube from London to Liverpool and an automated version of tick-tack-toe. Charles Babbage, the propellor-head in question, was one of the greatest mathematicians of his time and I suppose that's why one of the things he *did* find time to invent was shoes for walking on water. Most of his life, however, was taken up with trying to raise money to build two geared calculating machines of such complexity that he never built one that worked. One used punched cards and had stored programs and that's all I want to say about that. Ada, Countess Lovelace, said more than enough for all of us.

She was Babbage's aristocratic patron and promoter, and introduced him to all the right people, in return (they say) for Babbage providing her with a betting system. Like his machines, it never worked and caused a scandal. Rather like Ada's short-lived father, Lord Byron. Who spent a lot of his adulthood traveling in the eastern Mediterranean, where, in 1809, he met an odd cove called John Galt, out there trying to set up a Grand International Scam. At the time, Napoleon's continental blockade was ruining the U.K. export industry, so Galt's idea was to sneak British manufactures through Istanbul and into Europe by the back door, over the Hungarian border. Almost as soon as it started, Galt's entire shaky enterprise went down the toilet.

His major client, James Finlay, cotton manufacturer and wheeler-dealer of Glasgow, Scotland, took over and ended up briefly (till Napoleon's defeat) running a highly successful, Europe-wide network of blockade-runners. Finlay was pals with all the Industrial Revolution movers and shakers, including Richard Arkwright. In 1771 Arkwright's water frame had turned the cotton industry from out-work piecemeal into factory mass-production. A single power source (water) turned hundreds of rollers and spindles that pulled out the thread and then twisted and wound it, ready to be used on looms. Five years later came the patent on a single power source (steam) that would run Arkwright's machine and any other you could think of. Watt's steam engine was so popular that he couldn't keep up with the paperwork.

So he next invented a copying machine. Writing (or any design to be copied) was done on paper with a special ink whose ingredients included gum arabic. The completed original was then rolled against wet paper, on which the copy would appear (and last for twenty-four hours). In 1823 Cyrus P. Dalkin of Concord, Massachusetts, improved on the idea by coating one side of a sheet of paper with paraffin wax and carbon black. Pressing onto the sheet transferred a copy to the paper beneath. Dalkin called the product "carbon paper" and sold it to Associated Press. In 1868 AP sent a reporter to cover a balloon ascent by Lebbeus H. Rogers, who was a biscuit-maker. Well, these were eclectic times. Rogers was in the AP office doing an interview after the flight, when he saw Dalkin's paper at work. Instantly quitting biscuits and balloons, Rogers went into the carbon-paper business. In 1873, he went to a demonstration of the amazing new typewriter, where he persuaded the typist to try one of his carbon sheets. And the rest is history (unusually, repeating).

The typewriter Rogers saw had been manufactured by the E. Remington Company, because they had spare capacity and the kind of machine tools that would make the bits. There was little demand for the other bits Remington had previously made, once the Civil War had ended and there was a catastrophic drop in the demand for guns. The Remington had been one of the most successful guns ever made, rivaled in sales volume only by Sam Colt. Who made revolvers because his explosive mines failed. This may have been because in 1844, after he had successfully mined a ship on the Potomac, at a distance of five miles, he wouldn't give the Navy the secret, so they wouldn't give him the money. Immanuel Nobel was a great deal more open about *his* operations when the Russians asked him to make mines for them. By the time the Crimean War started, in 1853, "Colonel Ogarev's and Mr Nobel's Chartered Mechanical and Pig Iron Foundry" had been sowing mines everywhere around Russia for twelve years. One place they'd sewn up in this way was the harbor at Sevastopol. So the allied fleet supporting the troops in the Crimea was forced to anchor round the corner at Balaklava, where it was destroyed by the full force of the hurricane of November 13. A seven-thousand-ton cargo of medical supplies and clothing went to the bottom, leaving British troops to a terrible winter of pneumonia, starvation, and dysentery.

One week earlier, an extraordinary woman named Florence Nightingale had arrived in the Crimea. She and the thirty-eight nurses accompanying her spent that terrible winter discovering how bad British army medical services really were. She'd heard a few rumors: As the best means of warding off disease British military doctors recommended smoking, or growing a mustache (to filter the germs). In one recovery area, one thousand men suffering from diarrhea shared twenty chamberpots. In the hospitals the patients underwent surgery on floors covered with blood. Wounds were often not dressed for five weeks. The hospital mortality rate in most cases reached close to 50 percent. By the end of the war, of the 18,058 casualties suffered by the British, nine out of ten had died from disease. When these facts hit the newspapers back in England, everything hit the fan. Thanks to Nightingale's thousand-page report, filled with horrifying numbers, the Crimean War marked a turning point in military medicine.

Nightingale's obsession with statistics had started with a keen interest in botany. And it was while she was doing some botanical classification work that she had come across a statistical law that tickled her fancy and led her to strike up a lifelong friendship with its discoverer. It was Quételet's law about flowering lilac.

Hope all this has planted a few useful thoughts.

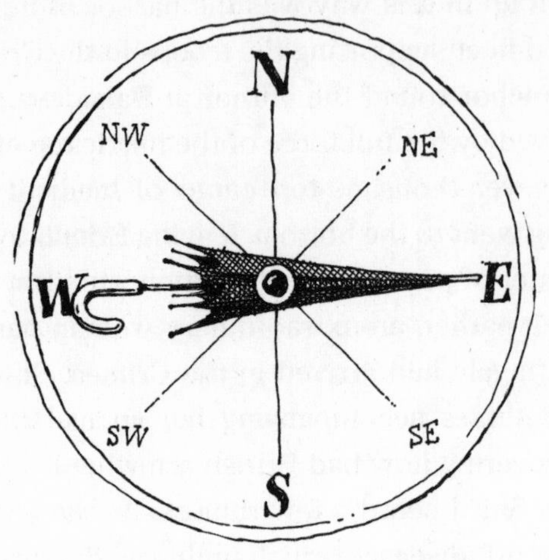

23

AND NOW THE WEATHER

IT WAS IRONIC that as I recently ran into the Paris church now known as the Pantheon, out of yet another rainstorm sweeping in from the Atlantic, I was filming a sequence about the guy who told us why the rain always comes from that direction. Because in the Pantheon hangs

Léon Foucault's great experiment of 1851, in which he dangled a sixty-two-pound cannonball on two hundred feet of piano wire, pulled it to one side with a cord, then burned the cord to release the ball without influencing its movement. Over the next few hours the pattern traced by the swing of the ball in the sand on the floor beneath the pendulum was the first physical proof that Copernicus had been right. As the pendulum swung in inertial space, a stylus attached under the ball made a line that shifted as the Earth turned beneath it. Point being, this demonstration would become the basis for the meteorological thinking of Buys Ballot (see elsewhere here) and others about how global weather was driven by the west-east rotation of the Earth. Made Foucault instantly famous.

Foucault's name went up in lights for other reasons too, however, thanks to his improved regulator that kept arc light carbon rods at the right distance apart as they burned. Made the light efficient enough to use in public places like theaters. Attended, in 1892, by people like pharmacist Henri Moissan on an evening off. Back in his lab Moissan was using the arc for a very different purpose: to power his electric arc furnace. In which the carbon rods burned so hot they nearly made him artificial diamonds. Moisson was *le noodleur extraordinaire* who got the Nobel in 1906, isolated fluorine (without killing himself) and wrote over three hundred scientific papers. In the winter of 1897 he also provided a young Polish physicist, living in Paris, with some uranium powder, and she began to investigate the powder's mysterious ability to "electrify" the air around it. This was Marie Sklodowska Curie's first step on the way to her discovery of what the "electrification" might turn out to be. Later she called it "radio-activity."

Marie had recently married into the Curie family, where

there was a strong tradition of egalitarianism, science, and emancipation. Much of this had come from the grandfather, Paul Curie, who early in the century had been a follower of New Christianity, a sect started by a weirdo named Henri de Saint-Simon who at one point, while temporarily destitute and starving, had shot himself in the head six times and survived. Saint-Simon's idea was to update religion so that it was closer to the modern world. The New Christianity prospectus talked about spiritual power belonging to men with practical knowledge and, in general, glorified work and the capitalist ethic. No surprise that by 1830 engineers, financiers, and businessmen were dressing up in New Christian costume and talking about free love (this particular aspect of the new religion would ultimately lead to its downfall).

One of these New-Age techies was somebody who made even Saint-Simon look normal: the corpulent banker Prosper Enfantin, who took over after Saint-Simon's death and became New Christianity's messiah. Enfantin modestly saw himself as half of Jesus Christ. The other half would be a soulmate, as yet to be identified. At one point, traveling to Egypt in search of The Bride (he never found her), he became excited about one of the bees Saint-Simon had had in his bonnet: building canals. Enfantin's 1847 survey for a Suez Canal (a vision he had about joining East and West) later gave him grounds for a dust-up with the influential Vicomte Ferdinand de Lesseps, when de Lesseps actually *built* the proposed canal.

Ferdinand was a diplomat whose father (as French consul for Napoleon) had taken an illiterate soldier and made him khedive of Egypt. So Ferdinand could do no wrong in Cairo. When he suggested the Suez Canal, it was just what the khedive was looking for to give his country a better reputation than dreadful. The canal took Ferdinand ten years and twenty-five thousand laborers, and he did the official opening in November 1869, at one of history's greater black-tie bashes. Glitterati wasn't the word. Dinner for eight thousand. Every one of whom cared about protocol and getting the seat they deserved. One of the few who didn't have to worry about these niceties was the wife of Napoleon III of France, Empress Eugénie. Who also happened to be Ferdinand's cousin.

Those few avid readers of my work will be aware that, as a small Spanish child, Eugénie had met and become great pals with the French novelist Prosper Mérimée, who charmed her rigid. And you know the old adage: Be nice to people on the way up. It paid off. Once Eugénie tied the imperial knot, Mérimée was flavor of the royal month and rapidly became *trés chic* among the chattering classes of Paris. And London. Where he got to know Antonio Panizzi, top bookworm of the British Library and megaschmoozer. If there was anybody worth knowing in England, Panizzi was on first names. Panizzi himself had run away to England at one point in 1823 when the Italian Secret Service got a mite too close for comfort after he'd been involved with some shady folk known as the Carbonari, who wanted all kinds of crazy stuff for Italy, like freedom of speech.

Panizzi's libertarian views (and his accent) went over very big in Liverpool, his first port of call, where he had been sent with an introduction to William Roscoe. Apart from being an extreme Italophile (me too—Italy is, in my opinion, something you catch, and for which, happily, there is no cure), and biographer of Lorenzo de' Medici, Roscoe was at various times banker, botanist, antislavery activist,

MP for Liverpool, publisher, and collector of rare books (and runaway liberals). In 1806 he also wrote a nursery classic: *The Butterfly's Ball and the Grasshopper's Feast*, which immediately took the fancy of the king and queen, and was then published by John Harris in his first, boffo-success series of children's books.

Harris was the successor to John Newbery, the first publisher in England to print illustrated books specially written for children at his London publishing house in St. Paul's Churchyard (where he also did people like Goldsmith and Johnson). In 1780 a list of Newbery copyrights included *Mother Goose*, translated from the French. Where it had been originally written by a man who also had the dubious distinction of keeping the books during the construction of Versailles. Charles Perrault was one of three brothers. Claude designed a bit of the Louvre and the Paris Observatory. Pierre (a tax collector caught with his hand in the till) founded the science of hydrology and worked with a guy named Edme Mariotte, who wrote Pierre's work up as his own. Mariotte managed to put up a few backs with this and other similar tricks. Huyghens accused him of plagiarism.

In the 1670s one of the things Mariotte did, which was almost undoubtedly his own, was to organize a chain of data stations across Europe from which he was able to put together a report on global and European winds.

And for the first time theorize what Foucault finally helped prove, at the beginning of this essay: that the weather moves west to east because of the way the Earth turns.

24

ON TRACK

I WAS ON a train the other day, sipping a cold beer in the bar and thinking how, when I was a kid, railway things used to go clackety-clack and now they just hum smoothly along. Continuous welded track, I believe it's called. Anyway, behind the bar on my train was a small fridge with "Linde" written on it. And I remembered reading somewhere that Carl Von Linde, one of the inventors of the refrigerator, had been put up to his chilly endeavors by the local brewers who wanted to keep their vats cool so that they could make their amber nectar in summer. This, while Linde was also working as a locomotive engineer. With instant coincidences like that, I thought, there had to be an essay here somewhere.

Sure enough, Linde triggered more than trains, fridges, and beer. In 1875 he started teaching at the Munich Polytechnic, and a year later he was inspiring a young student who'd had a lonely childhood and was fixated on fuel efficiency. Well, it takes all kinds. The young man in question did little about his fixation, while selling Linde's refrigerators, until 1897, when he came up with a wonder-engine that blew everybody away because it was reputed to work on anything from coal dust to peanut oil. And in petroleum-starved Europe, that was sweet music to anybody with haulage in mind. This turned out to be everybody from destroyer captains to farmers.

Which was why Rudolf Diesel became rich and world-famous overnight. So let's hear it for lonely childhoods and fixations. Much of Diesel's instant fortune came from the sale of distribution rights. One of which was for the British Empire, to an American who did so well by Britain they made him a Sir. Hiram Maxim gave the Brits the first successful automatic machine gun, which was then immediately adopted in one form or another by every major military power. Another case, after Diesel, of the world beating a path (believe it or not, Hiram *had* also devised a better mousetrap). Maxim's gun was the weapon that would change the world by removing hundreds of thousand of troops from it during World War I. Thanks to Hiram, WWI would become the event known to historians affectionately as the "machine-gun war." Nowhere was the slaughter more heroic than in the air, where the machine gun spawned a new comic book character: the fighter-pilot ace. None more celebrated than the Prussian daredevil aristocrat, Manfred von Richthofen, aka the Red Baron, who notched up eighty kills, ran a squadron known (long before Monty Python) as the Flying Circus, and was reputed to have said: "When I have shot down an Englishman, my hunting passion is satisfied for a quarter of an hour." In the end, Manfred won more medals than he could wear at one time, and was feted by his fellow-officers as the only man who could spin out of a dogfight upside-down and still know instantly which way was home.

In a manner of speaking, so could his grand-uncle Ferdinand. Well, he was a geographer. Wrote the first definitive study of China, in which he illustrated the effect of topography on the economy. Then went on to California, where he reported on the Comstock Lode. Then went back to Germany, where (thanks to friends in very high places) he was given the chair of geography at Leipzig. Ferdinand's real contribution to the sum of human knowledge was his invention of chorography (how an area is made up of many smaller units) and chorology (how these units inter-react with each other). Well, there was more to it than that, but this is just an essay.

Ferdinand's successor at Leipzig was guy named Ritter, who introduced humans into the geographical equation. Largely a spinoff from Romanticism, Ritter's interest in human geography (he invented it) originated in the man-and-nature relationship explicated by J. G. Herder, of whom I've spoken too many times before. Ritter was also profoundly influenced by a single meeting with adventurer-explorer Alexander von Humboldt, who'd just got back from five years in South America, where he'd climbed the Andes, found the magnetic equator and the Humboldt current, as well as the source of the River Orinoco, as well as carrying out hundreds of astronomical measurements, and have I said enough. Humboldt also gave maps the relieved look they have today. On his way home, Humboldt dropped by Monticello to visit his guru, Thomas Jefferson, who was, like him, an environmentalist before environmentalists.

To be fair, there was a bit more to Jefferson than ecology. Like being third president of the United States and designer of the Capitol building in Richmond, Virginia. Although some have said the credit for that should have gone to Charles-Louis Clerisseau, crack French draughtsman and pal of Jefferson's. Clerisseau never got his due from Robert Adam, either. Adam was the Scots architect who made Neoclassicism the lifestyle of the rich and famous (at least in Britain) and learned much of what he later turned to profit from Clerisseau, when the two of them spent a couple of years together in Italy and Dalmatia, sketching classical ruins and giving Adam his ideas. Back in Britain by 1758, Adam was so successful he was soon employing three thousand craftsmen on his various stately-home rehab jobs. His trick was to make your crumbling pile look like the Parthenon.

One of the craftsmen he hired was Matthew Boulton (later to become James Watt's partner), who specialized in ormolu and all kinds of metal frames. Boulton had a company that made everything from shoe buckles to sword hilts. This was why, in 1786, he knew what it took to design steam-driven coin-stamping machines, just at the time when counterfeiting had reached levels that worried the government enough to think about ordering replacement coinage. Boulton's machines could strike up to 120 coins a minute, depending on design complexity, and by 1792 he already had coinage contracts for the East India Company, the Sierra Leone Company, the American colonies, France, Bermuda, and Madras. Five years later the British Mint caved in and asked him to make their new two-penny, penny, half-penny, and farthing coins.

In 1817 coin design took a turn for the more elegant when the Mint brought in an Italian engraver, Benedetto Pistrucci, and he brought in a pantograph reducing machine.

Pistrucci first produced a large-scale cast-iron model of a new coin design, and then traced out the model's contours with the pantograph pointer set on one end of a rigid arm. A spinning cutter, set further down the arm, reproduced a scaled-down version of the design on a life-size die. Pistrucci used the pantograph to put St. George and the Dragon on the British sovereign and crown for the first time. The new classic look did not, however, help him get the job of chief engraver. He was, after all, an alien. A few years after Pistrucci's death, bitter and disappointed, coin-design models were being electroplated, and the dies for them made of steel, thanks to the work of William Roberts-Austen, new master of the Mint and alloy freak. One of his steel alloys became known as Austenite.

And when he'd finished making money making money, he went on to help make that train ride of mine (see above) as smooth as it was, with steel alloy railroad tracks.

Here's where I get off.

25 IS THERE ANYBODY THERE?

WE WERE HAVING a bit of harmless fun the other evening after dinner, doing a little table-rapping and glass-moving, and somebody suggested we have a go at seeing if Darwin was around. We did. He wasn't. The idea had come from somebody at the table who had been reading about the great unsolved mystery: Who came up with the idea of evolution first?

Was it Darwin, or Alfred Russel Wallace, the self-taught surveyor and insect freak (in six years in the Malay Archipelago he collected 125,660 species of creepy-crawly!)? In 1858, while in Borneo, Wallace sent Darwin some modest thoughts on the origin of species, and quicker than you could say "I thought of it first" Darwin had published The Great Book. There are those who have suggested . . . *you know what.* Perhaps we should have tried contacting Wallace instead. After all, he was a leading light in the spiritualist movement and said he'd never seen a medium who was a fraud.

This view was shared by other eminent scientists at the time, including hot-shot physics Prof. Oliver Lodge. Like Wallace, Lodge was particularly interested in thought transference. He's best known perhaps for his work on another, equally hard-to-detect form of message transmission—radio. To deal with which, he came up with the idea of a small tube of iron filings that would cohere when a tiny radio signal passed through them, and thus act as a detector. (I know somebody else did, too, and he appears elsewhere.) Lodge's gizmo paved the way for a Canadian, Aubrey Fessenden, who in 1906 succeeded in sending out continuous radio waves (as opposed to the intermittent signals Marconi and others were using), carrying voice messages.

At the news of this broadcast, the United Fruit Company went promptly bananas. Here was the perfect way to organize themselves so that their ships and trains got to the same place at the same time. These niceties matter in the banana trade, because bananas grow so fast you can have several harvests a year. So growers do. Fast growing means fast ripening, so getting the fruit to the consumer asap is advisable. I know all this because I've read a terminally boring tome by Alphonse Candolle, top banana in bananas back in the nineteenth century (OK, no more fruit jokes), who ran the Geneva Botanical Gardens after he took it over from his father.

Candolle *père* was pals with another Geneva boffin, name of Henri Saussure. This guy became a world-class geology star when his publications on the geological processes fostered the realization that the planet had been around for a little longer than (the up-to-then-official) five-thousand-odd years. This in turn helped lay the groundwork for the aforementioned Darwin. Since Saussure was now so famous, and since he was also nuts about Mont Blanc, the Swiss considered changing the name of the mountain in his honor. "Mont Saussure" didn't have quite the same ring, though, so they dropped the idea.

Anyway, Saussure had a favorite pupil, Aimé Argand, whom he introduced to the Paris science crowd, and by the autumn of 1783 Argand was busy helping the Montgolfiers launch their demo balloon flight for the National Academy of Sciences. A few weeks later the two brothers substituted humans for ducks and chickens, and the first manned balloon ascent took place. To the excited amazement of Benjamin Franklin, who promptly went back to the States and made a fuss about America needing a Shuttle before the Shuttle. As a result of which, nothing of note happened. Indeed, in France, Napoleon took against the whole idea of this *particular* use of French hot air, and so the nascent French Balloon Corps was disbanded. Bad news for one

Nicholas Conte, its instigator, who went off and invented a new pencil lead. But that's another story.

Meanwhile, back in the United States came the Civil War and a resurgence of interest in things aeronautic, in the person of Professor (for some odd reason American balloonists were given this academic title) Thaddeus Lowe. His chequered flying career reached its height (and so did he) on June 2, 1862, when he hovered two thousand feet above the battle of Chickahominy in his balloon *Enterprise* (at last, a Shuttle before the Shuttle). The *Times* of London reported that Lowe was able to report on every movement of the Confederate armies (below), to his Union boss (also below). This was achieved by means of a telegraph wire running down the anchoring rope to the ground.

The man who put Lowe up to this trick was George McClellan, General of the Army of the Potomac and a young whiz-kid, who saw the intelligence potential of what Lowe and his vehicle could do. The other intelligent thing McClellan did was to set up a Secret Service department for the Army, with the aid of an ex-barrelmaker turned private dick whom he had employed before the war, to keep an eye on the property of the Illinois Central Railroad, of which McClellan was prez at the time. If I also tell you that the railroad's lawyer was Abe Lincoln, you'll guess who this gumshoe was. Thanks to these friends in high places, sleuth Allan Pinkerton would go on to set up the country's most famous detective agency. It was Pinkerton who first recognized that crooks had MOs. He was also a master of disguises. And his casebook read like a Who's Who of the underworld, including Jesse James, Butch Cassidy, and the Sundance Kid.

But Pinkerton's most notorious effort involved a bunch

of Irish terrorists (or anarchists or radicals or whatever) called the Molly Maguires, who were operating in the Pennsylvania coalfields. These operations involved arson, general mayhem, and murder. Pinkerton decided to infiltrate the gang, and in 1873 sent in James McParlan, who had the right qualifications for the job. He was Irish, Catholic, and tough. In no time at all, McParlan was doing too well. That is to say, the Mollies liked him so much he was soon being invited to join their assassination squad. Desperate to avoid this, McParlan persuaded the Mollies that he was a drunk by taking to drink. Too effectively, since he became a hopeless alcoholic and would eventually die dried-out, in obscurity, in Denver. Thanks to his two years of weekly secret reports to Pinkerton, McParlan had fingered the Mollies well enough to lead to their capture and several executions.

McParlan's work was not, however, to go totally unrecognized. In 1914 he became the internationally acclaimed hero of a novel called *Valley of Fear*. Well, he *would* have been such, but for the fact that the author gave the book's detective protagonist a different name. McParlan's heroics had been appropriated by (the already internationally famous) Sherlock Holmes. And given McParlan's own fate, it's ironic that *Valley of Fear* was to be Sherlock's last case, too.

After which his creator, Sir Arthur Conan Doyle, turned to expressing himself through a different medium. The kind that sat around tables and got up to what I was playing at, the other night. Because in 1914 Doyle stopped writing and took over where Wallace and Lodge had left off: He became a leading light in the Society for Psychic Research.

Hope you found this column entrancing.

26 TURKISH DELIGHT

I WAS WATCHING TV one chilly evening recently and thinking about sun, sand, sea, and stuff when suddenly on the screen there was an ad extolling the tourist and cultural attractions of one of my favorite holiday spots: Turkey. One bit of the screen showed the tulip (almost the Turkish national emblem) and the other, the ruins of ancient Troy, the city that a German weirdo called Heinrich Schliemann "found" in 1872.

Schliemann was a self-made businessman who'd made a fortune in the California goldfields and then by marketing dyestuffs in Russia. At one point he became obsessed by the writings of Homer and decided to spend a fortune trying to prove that the *Iliad*, and Troy, and all that poetic rambling about Helen launching a thousand ships, had all really happened. He failed, but his work was to stimulate real archeologists to take a closer look, later on. Meanwhile, Schliemann's sidekick in these efforts was a medical genius (and fellow Homer freak) named Rudolph Virchow, known from his imperious mein as the "Pope of German Medicine," who, apart from anything else, virtually kicked off public health and is celebrated as the discoverer of cellular pathology.

It was Virchow who made the momentous statement that was to change medicine: *Omnis cellula a cellula* (all cells come from other cells). By identifying the cell as the ultimate unit of life and disease, Virchow also paved the way for chemotherapy. For all these reasons, Virchow provided Schliemann with what that egregious, bad-tempered conman and thief desperately needed: an aura of scientific respectability. But Virchow was also in Troy for his own reasons. He was an amateur anthropologist, and interested in the history of human culture.

Anthropology had been more or less invented in Germany by Johann Blumenbach, who, among other things, related skull shape to racial classification. He did so by placing a skull between his feet and looking down at it. This became known among his adherents as the "Blumenbach position." Using it, Blumenbach divided humans into five racial groups, to which he gave names, one of which is still in relatively general use: "Caucasian." In 1724 Blumenbach was asked to investigate the case of the "Savage Boy," an orphan child discovered in Hanover and said to be a living example of a prehistoric human. Blumenbach eventually demolished this argument, but not before the boy had been sent to London where he was cared for (and exhibited, and much discussed among philosophers) by the queen's physician, a gent I've mentioned elsewhere, Dr. John Arbuthnot, whose work on statistical probability galvanized a dull Dutchman called William 'sGravesende.

In 1736 this person (whose life is described even by ardent biographers as "uneventful") was teaching Newtonian science in Leyden University and had a visit from a Frenchman engaged in writing a general guide to the great English physicist's work. Thanks perhaps to 'sGravesende's advice, the Frenchman's book would make its author the most famous science writer in Europe. His name was Voltaire.

As it happens, Voltaire also knew Arbuthnot, since they'd met when the French thinker was in London, at which time he had gone with Arbuthnot to see the latest theatrical smash hit. This was John Gay's *The Beggar's Opera*, the first real lyric opera and a rumbustuous satirical swipe at the political establishment. Apart from getting its author into deep doo-doo with the authorities, Gay's effort was given boffo rating by the chattering classes and set box-office records by running for an unprecedented sixty-two performances.

Gay's work had originally been talent-spotted by the manager of Drury Lane Theater, John Rich. So when the reviews came out, it was said that the extraordinary success of the piece would make "Gay rich and Rich gay." Rich himself had an eye for what was going to make the grade with theater groupies, which was why he also staged the first real ballet, by a fellow called John Weaver, who gleaned what he knew about dance steps from the recently published translation of a French book on choreography. The original author had in turn snitched the material from the dance master to Louis XIV, Pierre Beauchamp, the first to formalize the basic five feet positions, who developed a sophisticated method for annotating dance movements, and introduced French ballet terminology, like jeté and pas de deux, still in use today. Not surprisingly, as dance master to the king, Beauchamp worked closely with the music master to the king: Jean Baptiste Lully, a devious Italian who'd changed his name, and who also wrote the first military band marches for the new French army.

The new army was new because it was the first-ever full-time professional standing one in Europe. This radical approach to military matters had been perfected by the French minister for war, the Marquis of Louvois, who realized that the megalomaniac maunderings of Louis XIV meant it was time to start making France great. This would involve doing things to make others *less* so. An army would help. Louvois also saw how the new flintlock musket and socket bayonet were going to turn warfare into something not so much hacking and yelling as more discipline and training, if the new weapons were to be used to fullest advantage. That is, with clockwork precision, by uniform lines of men, maneuvering and firing by numbers. Louvois' new concept of a permanent professional standing army finally did away with the traditional practice of using mercenary troops to fight one's battles.

This military re-engineering went over like a lead balloon with the traditional source of the best mercenaries in Europe: the Swiss. Principally because the technique that had made them so popular had been the pike square, in which large numbers of pikemen would protect small numbers of musketeers by standing round them with their twenty-foot, steep-pointed pikes, which they would lower into a kind of hedgehog formation any time enemy cavalry appeared. At which point the cavalry would stop and the musketeers would knock them off. The new musket and bayonet would do both jobs in one.

There was one bit of Switzerland that didn't care about this high-tech destruction of jobs: the canton of Zurich, where, some time back, they had already canceled their mercenary contracts, thanks to the activities of a holy-roller type named Ulrich Zwingli. By 1520 this firebrand religious reformer had effectively taken his community out of the Catholic Church with such un-Roman ideas as eating sausages in Lent, allowing priests to marry, and removing organs from churches, as well as taking down the statues and pictures of saints, discontinuing the mass, conducting services in German instead of Latin, prohibiting alcohol after sundown, and outlawing low-cut shoes. Some killjoy.

Zwingli's godson was an equally pious wimp: a scribbler named Conrad Gesner, who tickled Ulrich's fancy with products like the Lord's Prayer in twenty-two languages and a giant catalogue about all books ever printed to that date. One of the other things Gesner got up to was formulating a new (the first) classification of animals based on their physiology, and of plants (the first) based on their shape and seeds.

As part of this latter botanical effort, in 1576 Gesner also published a book containing the first European drawings of a knockout new flower, recently arrived from exotic foreign parts.

The tulip on my TV screen the other night.

27
SHEER POETRY

Give me your tired, your poor,
Your huddled masses yearning to breathe free

THE POEM ON the Statue of Liberty is a great reminder of how often the best-laid plans get hit by Murphy's Law.

In 1871 France had just lost a war to the Prussians and was beginning the later-familiar business of French governments' imitating cuckoo clocks (in and out, on the hour). In the politically unstable climate of the times, yo-yoing between monarchy, the Terror of the French Revolution, and moderate republicanism, the third in this group were keen to find some way of warding off the threat of a return to one of the other two.

So the French government thought up a gigantic statue, to be built by the French and dedicated to republican ideals, to be erected in New York Harbor (the gateway to the country whose independence the French had bankrolled a hundred years earlier), that would serve to remind any Frenchman who might want to go back to the bad old days of the Terror of the link between the two countries, and of France's natural but "moderate" republicanism. Good plan, but fifteen years later this crafty political ploy had been thwarted by the no-fool Americans. Thanks to Emma Lazarus's hymn to America, engraved on the statue's pedestal, the Statue of Liberty was, already at its inauguration in 1886, perceived to be less an acknowledgement of the debt to France and more an exclusively American statement of the country's open-door policy to immigrants (even to those escaping persecution in France).

The French engineer who built Liberty was Gustav Eiffel, the hottest monumental type around, with dozens of bridges and aqueducts to his name. A few years later he would realize his dream of building the highest tower in the world, using the lightweight, trussed wrought-iron structures he so successfully employed on the great statue. When he finished his tower, in 1889, it rose 986 feet above Paris and came in under budget. Ready for a lot of "blot-on-the-landscape" criticism, Eiffel designed the tower so as to be easily dismantled (it nearly was, in 1909, but in the new era of radio telegraphy the altitude of its antenna saved it).

Eiffel and many others took advantage of the tower's height to conduct experiments that might otherwise have required balloons. He dropped various airfoils and studied their falling behavior. The results were encouraging enough for him to set up a wind tunnel (it kicked off scientific aerodynamics) at the base of the tower. At one point, not long afterward, the president of the Aero Club de France carried out similar aerodynamic droppings, and then set up a ninety-foot manometer on the tower to test the pressures of various liquids and gases.

He was pressured into this activity by his search for extreme cold. Louis-Paul Cailletet ran his father's foundry and may have been looking for ways to provide a source of the oxygen needed for the new Bessemer steel-making process. If the oxygen were cold enough, you could store it in liquid form for use whenever it was needed (if you could find a way to store it). Anyway, in 1877 Cailletet managed to liquefy oxygen, using a technique that involved the drop in a gas temperature that follows a drop in its pressure. However, before he could report success, a cable arrived from Geneva. Sent by a refrigeration engineer, Raoul-Pierre Pictet, it claimed he'd done the same, only different. Pictet's approach had been to use a "cascade" process, in which a series of coolant gases, each of which liquefied at a progressively lower temperature, chilled the next, and finally made oxygen liquefy.

Soon afterward, a Scotsman named James Dewar, who had an obsession for getting to absolute zero, used these techniques, and 1898 succeeded in liquefying hydrogen. At 260 degrees Centigrade, solid hydrogen was within 14 degrees Centigrade of Dewar's frigid goal. One of the reasons Dewar did so well was his invention of an insulating jacket (which would solve Cailletet's problem with storing oxygen for steel-making) in which a vacuum between two silvered layers of steel or glass kept cold things cold. With this ability to prevent his chilly liquids from boiling away, Dewar's reputation for being the ultimate in cool attracted the attention of anybody who wanted to see how his pet project might do in the freezer. One of whom was Pierre Curie, of recent radium fame. Dewar then helped Curie investigate the behavior of radium and in particular the gases it would absorb at very low temperatures.

The way Pierre and Marie Curie had succeeded in finding radium in the first place was by dumping tons of pitchblende into vats, boiling it down, and then measuring various pitchblende characteristics. One of these was that the concentrate very slightly charged up the atmosphere immediately around it. *So* slightly as to be almost immeasurable. Until the Curies used the properties of a piezoelectric crystal to measure the charge. Piezoelectric crystals (such as quartz) react to even an infinitesimally small charge by changing their shape.

One of the Curies' most ardent supporters (ardent in more senses than one, as he was later to become Marie's lover) was Paul Langevin, who for many years assisted them in their lab. Later, Langevin investigated further the *other* thing piezoelectric crystals will do. When their shape is changed by pressure, they give off an electric charge. By 1917 he had produced what came to be known as the "Langevin sandwich." This consisted of a layer of quartz between two layers of steel. Zapping the quartz with electricity made it change shape a zillion times a second. This caused it to set up resonance. When placed in the hull of a ship, the outer steel layer transmitted a powerful resonating signal into the water (Langevin killed many fish in his early experimental water-tank tests). When the signal hit an enemy sub (or any solid object such as reefs or the bottom) it bounced back, setting up vibration in the steel plate. This caused the quartz crystal to resonate, producing an electric charge that made the familiar ping you hear in all the underwater war movies. Sonar.

The first discovery of this piezoelectric crystalline behavior was triggered around 1802 by the canon of Notre Dame cathedral in Paris, Rene-Just Hauy, who also succeeded in the baffling enterprise of becoming an associate member of the French Academy of Science *Botanical* Class, thanks to his paper on the crystal forms of garnet and Icelandic spar (you work it out). Hauy founded modern crystallography when he investigated why crystals seemed to shatter into uniform and identically shaped bits, after he hit them. As a result of this smashing performance, Hauy was loaded with honors and important positions, but lived a frugal life, spending all his money on support for the work of his brother Valentin.

Who in 1784 founded the first Institute for Blind Children in Paris. In 1826 Louis Braille became one of its teachers. Three years later he published the reading system used everywhere today: six dots arranged in two rows of three, offering sixty-three possible embossing combinations with which to express the alphabet, frequent words, punctuation, numbers, and so forth. A little later, the school was visited by Samuel Grindley Howe, who in 1832 became director of one of the first American educational institutions for the blind: the Perkins School in Boston.

It was Howe's wife, Julia, who would write the other great hymn for America, besides the one on Miss Liberty: the "Battle Hymn of the Republic."

28 LUCKY HE MISSED

I WAS IN the London Zoo the other day, staring at a buffalo, thinking about the fact that such zoological places all started as a "get-inside-God's-head" attempt to reproduce the two-by-two conditions on board Noah's Ark, thanks to the work of people like an obscure, middle-of-nowhere-

North-of-England country vicar named William Paley, who wowed everybody with the 1802 equivalent of chaos theory.

But Paley's was an *order* theory, and he explained it all in a large book entitled *Natural Theology.* What gripped the public imagination was his idea that every bit of nature was like a watch: designed with a purpose. Thus: Cranes can't swim because they haven't got webbed feet therefore they have long legs so that they can wade. So for early zookeepers, if you managed to collect all the animals in existence in one location, you'd get an idea of what He (they didn't think of Him as a She) was thinking at the time of Creation, and maybe work out the heavenly watchmaker's "Grand Design."

The guy who hoped to turn this theory into practice by setting up the London Zoological Society was a big fan of Paley, and while he was briefly British governor of Java had spent most of his time in the jungle, scooping up anything that walked, crawled, flew, or sat there long enough. Sir Stanford Raffles (for it was he) also deviously obtained a very long lease on Singapore for the Brits and instantly became one of the Great and the Good, back home. He got the London Zoo job in 1826 because of another one of the G & G fraternity. Name of Sir Humphrey Davy, who lobbied successfully for Raffles as Zoo prez.

Davy was about as big a science wheel as you could get: so eminent a savant that he was able to collect a medal from Napoleon's French Institute in spite of the minor inconvenience of Britain and France being in a state of war at the time. At the tender age of twenty-three Humphrey had made so much of an impression with his chemistry experiments he was offered the job of assistant lecturer at the Royal Institute in London. His first talks on galvanism (aka electricity) got him rave reviews and swooning ladies. By 1806 he was running the place and had become the hottest thing in electrochemistry. And since that kind of guy always knew everything, in 1812 when a mining disaster killed ninety-two people, Davy was approached to solve the problem of fire-damp. This explosive mixture of air and methane was often found underground, and if you happened to come across it with your lighted candle you tended to get taken seriously dead. In no time at all Davy had the answer in the form of a lamp whose flame was surrounded by a fine wire gauze. The flame burned, but the surrounding gases didn't.

As a result, Davy was awarded a humungous money prize by his pals in the Royal Society. Most unfortunately, an uneducated and unknown collieryman, who claimed to have done the same thing only better, wasn't. Fortunately this individual had other fish to fry. Mine owners, concerned at the way the Napoleonic War was causing the cost of horse feed to spiral, were desperate for alternative hauling power. So in 1829 our slighted lampmaker (George Stephenson) came up with a traveling steam-power gizmo called a "locomotive," and instantly became a railroad bigwig, feted by royalty everywhere. Better late than never. As were his trains.

George's son Robert picked up where his dad had left off and became a famous engineer, in 1850 opening his revolutionary Britannia Bridge, which joined England and Wales by means of two giant cast-iron tubes through which trains passed. The bridge probably was a Guinness record before Guinness (I don't drink the stuff, so I don't know when "before" was!). The record in question was no fewer than 2,190,000 metal studs, banged into holes put there by an ingenious automatic machine working on punch-card control. This struck Robert's friend Isambard Kingdom Brunel as an absolutely riveting idea, since he had a plan of his own that would need 3,000,000 of them (well, these people *were* Victorians).

In 1866 Brunel's plan, by this time known as the SS *Great Eastern*, biggest ship in the known universe, was inching into Heart's Content Bay, Newfoundland, hauling one end of the first successful transatlantic telegraph cable (the other end was anchored to Valencia Island, Ireland) and making the day for a certain Cyrus Field. Who was a retired American millionaire papermaker, and he owned all twenty-five hundred miles of the cable (and a further thousand miles of another one, broken off earlier and now at the bottom of the sea), so the Morse Code now coming ashore in Newfoundland was music to his . .- .-. ...

Morse himself was one of Field's advisers, having had experience in laying cable ever since 1844 when he had done it between Baltimore and Washington, so as to transmit the first telegraph message, "What hath God wrought," and stupefy Congress. Not stupefied enough, though, to finance his idea. Fortunately his business manager was an astute type named Amos Kendall, a former U.S. Postmaster-General, over whose land Morse's Baltimore–Washington cable had crossed. It was Kendall who suggested to Morse that he'd do better setting up a private telegraph company instead of pressing for government support. In return for this blindingly obvious idea Kendall got 10 percent of the first one hundred thousand dollars Morse would make and 50 percent of the rest. So by 1864, guess what, he was a very rich man. Since he was married to a deaf woman (as was Morse), Kendall decided to give some of his well-gotten gains to help found the first National Deaf Mute College (now Gallaudet University).

The mid–nineteenth century witnessed great deal of American interest in speech and hearing impediments, as well as arguments about how they should be treated. Several schools for stammerers were established by another self-made man who'd also made a fortune in communications. This was William Fargo, a stammerer himself, who'd started life as a freight agent in New York and had gone on to partnership with Henry Wells in a courier company they set up in 1850, called American Express. That year, fifty-five thousand people had gone west to California. Thirty-six thousand of them had made the trip by sea, as did most of the mail, since dying of thirst or sunstroke, and the objections placed in your way by native Americans, all tended to make the overland route somewhat iffy.

As ever, it was money that would surmount these minor inconveniences. In 1858 gold was found in Colorado and Kansas, and two years later the miners were getting their letters hand-delivered by individuals lathered in sweat and covered in dust because they'd done the last hundred miles at full gallop. Wells and Fargo ran the western end of this extravagant and brief-lived door-to-door service, known as the Pony Express. Extravagant because it lost a huge amount of money, and brief because eighteen months after it started it stopped, when, in November 1861, the coast-to-coast telegraph link was completed.

By that time, however, one particular rider had already left to provide meat for the Kansas Pacific Railroad, because apart from his ability to ride hell-for-leather, he was also a crack shot. Well . . . fortunately . . . not absolutely crack. Although his record kill of 4,280 animals in eighteen months (69 in one day alone) was so impressive it earned him his nickname, "Buffalo" Bill Cody can't have been that good, or I wouldn't have been able to admire that magnificent beast the other day in London Zoo.

Well, I must shuffle off.

29
CHEERS

A RECENT BARMAN opened my tonic water bottle with a flourish and the tinkle of metal reminded me of William Painter, the man who invented the Crown Seal Company bottle cap (and then blew his Hall-of-Fame chances by advising one of his salesmen, name of Gillette, to invent a similar use-and-throw-away gizmo whose reputation would totally eclipse that of bottle caps).

Anyway, after many earlier closure attempts including glass balls, wire and cork, or wax, capped effervescence was first made generally available by Jacob Schweppes at the 1851 Crystal Palace Exhibition in London, when he sold vast quantities of soft drinks, thus realizing the dream of long-dead Joseph Priestley, who'd invented the soda water decades before. Getting fizz into water (so that, as was thought, it would cure yellow fever) was one of Priestley's more successful industrial-chemistry efforts, all inspired by the modern education he'd received at one of the great Dissenter academies.

These had originally been set up in late-seventeenth-century England by Protestants who wouldn't accept the return of a monarchy after the failure of Cromwell's Puritan Commonwealth. The reason these guys wanted to open their own schools was that, as a result of refusing to sign an oath of allegiance to the (Catholic) monarch, those who had "dissented" in this way were barred from going to university, becoming members of Parliament, preaching or trading in the major cities, and taking jobs in the army. So their options for advancement were somewhat limited.

The up-to-date curriculum of the Dissenter academies included unheard-of subjects such as science and modern languages and had been generally influenced by the ideas of a Czech free-thinker and educationalist named Amos Komensky. This Bohemian theologian had arrived in England in 1641 and made such an impression on the Puritans with his two books on pedagogy—*The Great Didactic* and *The School of Infancy*—that he was showered with job offers, one of which was to take over the presidency of some New England college nobody'd ever heard of, called Harvard.

Komensky turned the post down, preferring to concentrate on the further development of his philosophical views. These included thoughts about reality and how it was made up of irreducibly small elements, a concept said to have inspired a certain German math freak named Gottfried Leibnitz to conceive of fundamental entities that he called "monads." And then, in 1675, to invent (or not, if you're English and believe it was all done by Newton) the calculus you needed in order to measure infinitesimally small matters.

The other (and larger) thing Lebnitz was crazy about was libraries. His job as book keeper to the Elector of Hanover was pretty much of a sinecure, given him so he could write (he never finished it) a history of the ducal family. While book-hunting in Paris, Leibnitz was apparently much taken by a text on how to put together your own library, with instructions on how to catalogue, choose titles, dust books, and treat the library staff. This bookworms' delight had been written in 1644 by Gabriel Naudé, who'd put his money where his mouth was by collecting and organizing a gigantic library of forty thousand volumes for his boss Cardinal Mazarin, first minister of France, who then built a place to house it all, and opened it to the public.

Naude's book also caught the eye of an English aristo and scholar, John Evelyn, who came across it during a tour of Europe the following year. Evelyn eventually translated the book and gave a copy to a friend, who used it to organize the pile of material he had amassed (he chose books by their size rather than subject matter, except in the case of erotica) as part of a great plan to write a history of the English Navy. It was only thanks to this buy-it-by-the-yard bibliophile, Samuel Pepys, that in 1688 there was anything naval to write about. As secretary to the Admiralty, it was Pepys who made sure Britannia was able to rule the waves, by introducing standardized ordnance, regular shipbuilding programs, official rates of pay and promotion, disciplinary codes, pensions, and a new breed of naval captain who knew bow from stern. About the only reformatory matter at which Pepys was a signal failure was signaling. Which at the time wasn't up to much.

The contemporary limitations on what you could say with a few limp flags on a windless day is best illustrated by the fact that even in a gale, if the admiral were inviting you to lunch, the flagship hung up a tablecloth; if they wanted wood, an axe. By 1794 things had improved a little, but from the point of view of the British government, the real problem was that the Navy was still not getting the message fast enough. Specially between London and the Fleet headquarters in Portsmouth. So that same year, when a French prisoner-of-war was discovered to be carrying a copy of instructions for a radically new semaphore communications system (recently invented by M. Claude Chappe, and already in use by Britain's deadly foe, Napoleon), suggestions for an improved version were instantly forwarded to the authorities by Reverend John Gamble, the military chaplain who'd come across the information. Gamble's semaphore used a wooden frame carrying five shutters that could be opened and closed in coded patterns that could be seen by telescope some distance away. A chain of stations to relay these patterns would be able to get messages from Portsmouth to London in a matter of minutes. Unfortunately, a further, minor improvement was proposed by another clergyman who also happened to be the fourth son of an earl, so commoner Gamble's idea bit the dust. And he went back to buying foreign patents. One of which was for a French food-preserving process.

By 1818 the most recent attempt to search the polar regions of Canada for the Northwest Passage set sail with provisions of the new canned food (the expedition also carried forty umbrellas as presents for Eskimos). Ultimately, the enterprise failed in its quest, but the experience whetted the appetite of James Clark Ross, on-board nephew of the expedition leader, and in 1829 he headed his own venture: to find the Magnetic North Pole in the same approximate area. On the morning of June 1, 1831, when Ross hung a magnetic needle on a fine thread of New Zealand flax, the needle's dip of 89° 39′ was close enough to vertical to convince Ross that the Pole was underfoot. The location, at 70° 3′ 17″ N 96° 46′ 43″ W, was marked with a cairn of stones, a flag was flown, and the Magnetic North Pole was claimed in the name of Great Britain and King William IV (Ross was unaware that even as he spoke, the wandering magnetic spot was already on its way elsewhere).

As is often customary with explorers, Ross gave names to some of the various desolate places he got stuck in. On this occasion these names included the Boothia Peninsula, the Gulf of Boothia, and Felix Harbor. You'll gather from this, that somebody called Felix Booth was worth commemorating. In fact, had it not been for Booth's generous gift of twenty thousand pounds, Ross's entire voyage might not have been possible, the Magnetic North Pole might not have been (temporarily) British, and I might not have managed to make this connective tale end the same way all these columns do: back where it began.

In that bar, remember? Where the *other* thing the barman was pouring into my glass, besides tonic, was what made Felix Booth rich enough to finance polar expeditions. Booth's Gin.

30

WHAT'S IN A NAME?

NOT LONG AGO, while I was wandering through that treasure house of technological history, the Smithsonian Institution in Washington, D.C., I was reminded that evolution seems to have made us the only animal on the planet with a conscious appreciation of its own past.

Which may be why, in 1801, James Macie, scientific dabbler and the illegitimate son of the duke of Northumberland, took on the aristocratic family name after his father's death meant that there was nobody left to prevent him from doing so. Macie had two notable claims to fame: He (a) wrote one of those monographs-you-can't-pick-up about a strange kind of bamboo-joint juice called tabasheer, and (b) devised a new and improved way to make good coffee. That he was elected to the harrumph Royal Society may have had more to do with the influence of his immensely rich scientific (and noble) pal, Lord Henry Cavendish, discoverer of hydrogen, who took Macie under his wing (noblesse oblige—everybody knew Macie was a bastard, but nobody ever came out and said so), and gave him free run of his private lab in London.

Cavendish himself was a history fetishist (he dressed in clothes of his grandfather's time); he was described as "bashful to a degree bordering on disease"; and he went around uttering sudden loud squawks. He also became enmeshed in one of those "who did it first?" rows about the composition of water. It happened because the Royal Society got the date of Cavendish's 1783 paper, *Experiments with Air*, wrong by a year (late). And since everybody and his dog was investigating the same thing at the same time, the Society's error led to charges that Cavendish had plagiarized the whole thing from a similar paper by James Watt. In the end Cavendish and Watt settled the dispute amicably over dinner at the Royal Society and Watt went back to his steam engines, at the Birmingham factory that he shared with partner Matthew Boulton.

In 1779 Boulton had taken on a job applicant called William Murdock, who made an impression at the interview by dropping his hat. When Boulton expressed surprise at the resulting noise, Murdock explained that the hat was made of wood and that he had turned it on a lathe. Boulton made a good choice. Murdock went on to invent the "sun-and-planet" gearing system that transformed the back-and-forth thrusts of Watt's steam pump shafts into the rotary motion that would drive the wheels of the Industrial Revolution. And then he made these revolutions (and the Revolution) work a little harder by making it possible for everything to go into the night. In 1803, thanks to a process with which Murdock had been experimenting for eleven years (and which he had snitched: see elsewhere), the Boulton-Watt factory became the first industrial premises to be lit by Murdock's amazing, if rather smelly, new coal-gas.

Well, not that amazing to a Heidelberg chemist who, forty-nine years later, wanted a flame that would be free of all those malodorous impurities. Robert Bunsen was fascinated by (and assiduously visited) things that gave off fumes, like volcanoes and geysers in Iceland, or factory chimneys in England and Germany. Bunsen was especially concerned with devising ways to recycle the heat being expensively lost up the flues at iron foundries. So he was hot stuff on hot stuff. Which is why his name is familiar to any school kid who has ever done any lab work. The burner he invented produced a nonluminous coal-gas flame, free of anything except what you chose to burn in it. In fact, Bunsen chose a great number of things, aided by his sidekick, Gustav Kirchoff, who gave him the idea.

As ever, Kirchoff got the idea from somebody else: a glassmaker named Joseph von Frauenhofer, who some decades earlier had been checking his glass for imperfections by looking though it at the fine dark lines he saw in the rainbow of colors that appeared when he passed sunlight through a prism. Imperfections in his glass were easier to see like this, because they made these lines wavy or smeared. Turned on by the dark "Frauenhofer lines," Kirchoff and Bunsen chose the Bunsen burner as their light source, and started burning everything and looking at the light through prisms. Today we call what the two men were up to "spectroscopy." You pass the light from a burning material through a prism, and you see black lines in the spectrum at a set of frequencies (colors) unique to the stuff being burned. Look up the lines in your tables, and you know what the material is. If it's not there, you've found something new. And all you need to do this trick is a tiny amount of the material you want to identify.

In 1864, this last fact excited a Brit called Henry Sorby, who took his mother to every scientific conference and on all his expeditions, and who was a freak for the very small. Sorby had pioneered the technique of slicing rocks so thin that you could read a newspaper through them, and then submitting their structure to microscopic examination, to see how the rocks had been formed. (The shape of tiny cavities or bubbles, for instance, would indicate whether the rock had come into existence as the result of heat or pressure.) And because some of his samples were primeval, Sorby was able to say meaningful things about the earth's ancient past. As soon as he found out what was going on at Bunsen's lab in Heidelberg, Sorby broadened his field (by a zillionth of an inch or so), by sticking a spectroscope on the end of his microscope and analyzing the microconstituents of everything from poison chemicals to autumn leaves. It was while peering at the latter that Sorby found what makes them go russet: carotene, the pigment responsible for the vivid coloration of nearly every red-yellow-orange living thing.

In 1876 Franz Boll, a German studying frog retinas in Rome, came across that same material when looking for the visual pigment that enables the eye to see in both bright and dim light. Boll found that bright light bleached light-sensitive "rods" in the retina from red-purple to orange and then to white. Further examination revealed that the substance that reverses the process when the light level drops again was a form of carotene (lack of which was to explain night blindness and start a World War II ritual in which bomber pilots would eat a lot of carrots before night raids). Meanwhile, Boll visited Berlin, where he explained his work to various scientific-establishment godfathers, including Ernst Pringsheim.

Radiation physics was Pringsheim's particular line, and infrared radiation in *very* particular. For the investigation of infrared rays, he developed a special version of the radiometer, an instrument designed for measuring radiant energy. This device was to prove scientifically awkward when it was revealed to be not what everybody thought. The radiometer (today a toy known as a light-mill, often encountered in curiosity shops) had been invented by the eminent Victorian sage and experimenter William Crookes. It consisted of four tiny vanes, lampblacked on one side, attached to crossed arms that were delicately balanced atop a steel spindle resting in a cup; the whole was encased in a glass vessel that had been pumped out to a high vacuum. When a light was brought close to the little gizmo, it revolved. Crookes (and most others, including Pringsheim) attributed this to the impact of particles of light on the vanes. Wrong. As was embarrassingly revealed by a British flow expert, Osborne Reynolds. What was really happening, he said, was that the blackened side of the vanes was heating up, causing tiny amounts of gas trapped in the pith to expand and leak out. It was this escaping gas, and not the light, that was pushing the vanes around. Infrared faces all round.

Crookes, unperturbed, was busy on other inquiries. Apart from inventing the cathode ray tube and being elected president of the Royal Society, Crookes also hired himself out as a freelance chemistry consultant. As you might expect, given the ambiguities of Victorian culture, Crookes also had another, less public side to his character. He got his kicks from a ghost, name of Katie King. This lady would appear at seances, where she was photographed arm-in-arm with the infatuated inventor (who also frequently witnessed other paranormalities, such as self-playing accordions, levitating water jugs, and psychokinetic furniture).

Crookes's fascination with spiritualism was shared by many of his contemporaries, including a young man whose writings on the subject were quite well-known. Less so was his other principal area of interest. Alfred Russel Wallace spent some years in the Malaysian archipelago where, among other things, he identified a peculiar geographical dividing line to the east and west of which there appeared to be species entirely unrelated to each other. As part of his investigations into this unusual phenomenon, Wallace formulated a view of nature that would have put him on the scientific front page but for the deferential nature of his character. He let himself be persuaded to allow another naturalist, who had come to the same conclusions, to read a paper in their joint names at the London Geological Society, and then publish the joint thesis as a book. The publication shook the world of natural history to its roots. But there was only one author's name on it.

Which is why my opening Smithsonian thoughts about evolution were "Darwinian" and not "Wallacean." And also why I kicked off talking about James Macie. He was the fellow who left the bequest that founded the Smithsonian, and forever recorded his proud sense of history.

Because the family name Macie adopted after his ducal father's death was "Smithson."

31
FEATHERED FRIENDS

READING JOHN KEATS'S "Ode to a Nightingale" the other evening (well, why not) I was reminded of how nobody ever told us when I was at school in England that the English didn't invent Romanticism. That it all really started in late-eighteenth-century Germany with a group of scientist-philosopher-scribblers in Weimar, most of whom had "domestic problems." People like Goethe, Schelling, and the brothers August and Friedrich Schlegel. It was August (whose wife was fancied by August's brother, till she ran away with Schelling) who formalized the rules of Romanticism for everybody joining the new touchy-feely movement, including Keats.

Anyway, in 1804 August had the misfortune to fall for the salon queen of the chattering classes, Madame Germaine de Staël, famous for low-cut necklines, opinionated views on almost anything, and being on the run from the French police. Poor old Schlegel was to spend the rest of her life chasing around Europe being Madame's lap-dog and wishing she didn't have all those lovers. In between whom she managed to write a major work on German culture, pen a novel of "experience," put Jane Austen's nose out of joint, and make an enemy of Napoleon with some caustic comments on emperors. This was why she was on the run from his national security services. It's ironic that he should have taken quite so much against her, since it had been largely due to the inability of her father, Jacques Necker, to manage the French budget when he was the country's finance minister that had led to the chaotic events that brought Napoleon to power in the first place.

In 1778, before the French went revolutionary, Necker got a request from a Swiss inventor named Aimé Argand, who had a new process for distilling brandy and wanted a monopoly in southern France in return for making the technique publicly available. Necker agreed, and by 1782 Argand had three distilleries up and running. And problems with night shifts. So, being an inventor, he invented a light, on which I have offered illumination in another essay. Anyway, given the fact that contemporary Industrial Revolution England was where everything was happening, technology- and market-wise, by 1784 Argand was getting his lamps made there by a guy named Matthew Boulton. This canny type ran a factory in Birmingham, turning out everything from Argand lamps to shoe buckles to medallions to steam pumps.

Boulton's partner was James Watt, whose steam pumps were so successful because they worked so well thanks to the help he got (when he was Glasgow University repairman) from the chemistry faculty Prof, Joseph Black. Who'd done some whiskey-making experiments and discovered latent heat, thus facilitating Watt's idea of the separate condenser that made his steam pump go better than others. One of Black's other protégés was a medical student called James Graham, who went on to fame and fortune with quack electrical cures at his Temple of Health in London, where he employed a young ex-hooker called Emma Lyon.

After passing through the clutches of various aristo rakes, Emma eventually ended up as mistress and then wife of Sir William Hamilton, British minister to the Court of Naples. It was in Italy that Sir William started "collecting" antique vases and stuff found in the newly excavated ruins of nearby Pompeii and Herculaneum. From time to time he'd amass enough of this bootleg Roman crockery and broken Greek heads to put together a catalogue for discerning and well-heeled buyers in London. It was one of these picture books of Classical bric-a-brac that inspired the English potter Josiah Wedgwood to design his famous Neoclassical dinner sets used by queens and for good publicity reasons called "queen's ware."

On bright moonlit nights Wedgwood was one of a crowd of liberal thinkers, Freemasons, Quakers, and the like, who'd travel hill and dale to sit around in somebody's house and discuss everything from science to the latest doings of the rebels in America. One of Wedgwood's colleagues at these "Lunatic Society" meetings (they met at full moon) was a certain Joseph Priestley, who was, in time, to suffer for his support of the Americans by having his lab burned down by a mob. Shortly thereafter, in 1798, Priestley found himself across the Atlantic, among the Yankees he so admired. And (as one of the very few eminent European scientists to have done such an emigratory thing) being wined and dined on all sides. On one occasion by the faculty at Yale, where he bumped into, and impressed the pants off, a nervous young chemistry professor called Benjamin Silliman. Being somewhat of a hypochondriac, Silliman was into "curative medicaments" and stuff, so Priestley (inventor of soda water) was an instant role model.

The fact that soda water was supposed to remediate all known disorders encouraged Silliman into a disastrous soda-fountain venture in New York. Fortunately it was his mother's money. Later on, the Silliman family made up for this pseudoscientific blooper when, in 1855, Benjamin Silliman, Jr., analyzed some black sludge oozing from a creek in Pennsylvania and pronounced it to be "rock oil," later to become known as petroleum. There was only one problem with this amazing new energy source. What was the source? And how could you find more? Fortunately for would-be oil barons, the answer was at hand in the form of the work of somebody whose stultifying singlemindedness had already burst upon an astonished world back in 1826.

Alcide D'Orbigny had spent the previous seven years preparing a large work on a small subject (six hundred different species of a fossil marine micro-organism called *foraminiferum*). These little jobbies range in size from the supermicroscopic 0.01 millimeters to the gigantic 100 millimeters. And the profitable thing about *foraminifera* is that when two or three dead ones are gathered together, so to speak, there's likely to be oil somewhere close by. That's partly why much oil exploration consists of people getting eyestrain.

D'Orbigny was born in Santo Domingo and, after his family moved back to France, grew up in the small Loire village of Coueron. One of his boyhood pals, who also lived in both places, was a chap who shared Alcide's passion for natural history, and the two of them would spend many happy days on the banks of the Loire hunting for birds' eggs. Since the other guy also had a talent for drawing, and was almost as crazy about live birds as D'Orbigny was about dead bugs, he started to sketch (and then to paint watercolors of) almost anything with plumage. After he moved to the United States, this childhood hobby would end up making him world-famous among nineteenth-century naturalists. And equally well-known today to anybody who's ever got out of bed on a raw winter's morning to count migrants on the wing.

The fact that our painter was no good at figures may account for why, early in his career, he was such a disaster when it came to investment. At one point, in 1812 or so, before he hit the jackpot with his pictures, J. J. Audubon was living in Louisiana and got involved with a venture to run steamboats on the Mississippi. The whole speculative schemozzle ran aground and took with it the life savings of a couple of young English immigrants named George and Georgiana who'd come to America shortly after their wedding. I don't know what happened to the newlywed newly bankrupts, but the husband's famous brother wrote to them from England saying, in effect: "Audubon had better hope he and I don't ever meet on a dark night."

Guess who the brother was (and this is one of the greater examples of history's connections coming home to roost). John Keats.

Must stop now. Gotta fly.

32
SCRIBBLE, SCRIBBLE

One of the real pains caused by the kind of historical research I get involved with is that when you go to primary sources, if they lived back before the typewriter was invented a lot of the time you have to read somebody's scribble. You know: research notes, letters, diaries, that stuff. And most of the time, comprehension-wise, double-Dutch.

Recently, in my case, just that. Which is why what little I know about a nineteenth-century scientist from Holland named Christoph Buys Ballot is strictly secondary-source material. In more senses than one. Because I wasn't particularly interested in his primary work, which was to produce the law that says the angle between the wind and the pressure gradient is a right angle. What I found a lot more intriguing involved Buys Ballot riding with a trainload of brass players, on a railroad out of Utrecht, Holland, in 1845. The task of these traveling trumpeters was to blow a steady note as the train approached and then passed a group of Buys Ballot's pals. Who confirmed that they had heard the note rise as the train neared them, and then fall as it moved away down the track. Just as an obscure Austrian professor of math had predicted, three years earlier, would happen.

In fact, this Austrian's scientific paper had principally been about how the light from approaching and receding stars moved, respectively, toward the blue and red ends of the spectrum, as the perceived frequency of the light rose (approaching star) or fell (receding star). This phenomenon became known (after the now-no-longer-obscure Austrian) as the Doppler effect. And in 1859, it was turned to practical use when German experimental mavens Kirchoff and Bunsen (he of the burner) used their new spectrograph to show that you could use the shift in the spectral lines of the light from these moving stars to measure their velocity. Hot stuff. Unlike the obsession of one of their cooler lab students, who spent the winter of 1871 with them in Heidelberg.

Kamerlingh Onnes was his name (another Dutchman), and he was eventually to discover the strange things that happened to things when they got very cold. Things like superconductivity, which he discovered by chilling materials with liquid gases to within a few degrees of absolute zero. Onnes was aided in his endeavors by the work of a Frenchman, Louis Paul Cailletet (a fellow I have mentioned before), who in 1877 succeeded in liquefying oxygen. Cailletet was also an aeronautics buff and at one point invented a new breathing mask connected to a container of liquid oxygen so that balloonists at high altitudes could get high. In more senses than one. For which Cailletet was made life president of the Aero Club de France.

As it happens, the effect of sniffing excessive amounts of pure oxygen was the experimental concern of one of Cailletet's contemporaries, French physiologist Paul Bert. Bert, also known for his seminal work on how the sensitive mimosa reacts to touch (try it), spent many happy hours in a pressure chamber of his own design, finding out about stuff like the bends, hypoxia, and other such mind- and body-altering experiences. From time to time his happiness might possibly have been due to the fact that he was also trying to find out what pressures of different gases you needed in order to provide just the right amount of anesthesia for a surgical patient. One of these gases was nitrous oxide, aka "laughing gas."

The bible on NO_2 had been written back in 1800 by Humphry Davy, while he was working at the Clifton Pneumatic Institute in the west of England. Davy was soon to become England's premier scientist, a Sir, role model to Mary Shelley of *Frankenstein* fame (she wrote it), pal of the great and the good, and winner of all known honors. Nitrous oxide made Davy a bit of a laughingstock among his colleagues (in the nicest sense of the term), at the newly fashionable NO_2 parties, where researchers would get stoned, strictly in the interests of science. Alas, on a (slightly) less hilarious note, Davy failed to prove that the gas would cure all diseases.

As is often the case with scientists back in those Romantic Movement days, Davy also dabbled in a bit of poetry. About which little further need be said (read some). Mind you, when he became a famous lecturer at the Royal Institution, admirers of his versification included such Romantic types as Samuel Taylor Coleridge and Robert Southey, a couple of scribblers Davy had met during his time in Clifton.

In 1819 Southey spent some time touring the Highlands of Scotland with another eminent self-made person, Thomas Telford. This guy built more roads and bridges than anybody this side of the Romans, and especially two monumental efforts: one of the greatest engineering achievements of all time (so say the Welsh), the unparalleled aqueduct at the unpronouncible Pontcysyllte; and the *other* greatest engineering achievement of all time (so say the Scots), the Caledonian Canal (about which Southey enthused poetically). In a life of success piled upon success (well, structure was Telford's thing), the only failure of his entire career was in 1800, when he designed a bridge to cross the Thames in London. The top engineers and scientists of the time thought it was a winner. Alas, a reality check revealed that the government of the day could not afford to buy the land on either side of the river on which the bridge approach ramps would have to be built. So it was thanks but no thanks.

Among the regretful members of the Bridge Commission that turned Telford down was a young genius called Thomas Young, who was one of those people you love to hate. He knew every known ancient biblical language, cracked the mystery of Egyptian hieroglyphics, lectured at the Royal Institution on you name it, and conducted the famous experiment that seemed to prove that light beams from different sources produced interference patterns because light traveled in waves. OK, no problem. But . . . waves traveling *in what* ???

This mysterious lumeniferous medium became known as "ether," and the search for it bedeviled Victorian science as did little else. Hermann von Helmholtz, the harrumphest German scientist of the day, delegated this minor matter to a pupil (Heinrich Hertz, whose investigations, to see if electromagnetic radiation *also* moved through ether in waves, made possible the invention of radio).

Helmholtz himself had bigger targets to aim at, namely nailing the prevalent "Vitalist" theory, which held that the life processes involved some kind of immeasurable "force." Which Helmholtz disproved by showing that the passage of a signal down a frog's sciatic nerve was an entirely measurable twenty-seven meters a second. However, the etymologically implicit thing about a Vitalist theory has to be its ability to survive, right? So in 1900 Vitalism was still alive and kicking, thanks above all to one Ludwig Klages, German founder of the science of "characterology" (if they didn't offer it at *your* institute of higher learning, don't blame me). The other thing Klages did (which endeared him to the up-and-coming National Socialists a little too much for his taste, whereupon he left for Switzerland) was to develop a technique for analyzing character so accurately that, in spite of Klages, the Nazis would use it in their officer selection programs.

I could have done with Klages's trick when I was trying, at the beginning of this column, to make head or tail (it could have been "tail or head") of the scribblings of Christoph Buys Ballot. Because Klages's key contribution to the sum of human knowledge was to systematize the study of handwriting.

Must sign off now.

33 HEAVY STUFF

HINDSIGHT ALWAYS GIVES you 20/20 vision, doesn't it? So you wonder why, in 1609, when Johannes Kepler, an Austrian astronomer with the particularly unfortunate affliction of multiple vision, published his discovery of the mystery attraction holding the planets in orbit around the Sun (derived from all those numbers, mostly other people's data, given his eyesight), he didn't come up with a more scientific name for what the mystery attraction might happen to be. Nope. He called it: "Holy Spirit Force."

A couple of years later, when he met the English poet John Donne, Kepler gave him a copy of the book he'd written about his discovery, to take back to the king of England. We only know this (since the book never actually turned up in the royal library) because Donne's biographer and parishoner, Isaac Walton, says so. This was the guy who put an entirely new cast on fishing with something he wrote, in 1653, for all those out-of-work Anglican churchmen who'd lost their jobs during the Puritan Commonwealth, titled *The Compleat Angler: The Art of Recreation.* Actually the thing wasn't entirely his. The chapter on fly-fishing was written by a well-heeled literary pal and drinking buddy named Charles Cotton. Who also dabbled in a bit of translation. Cotton's 1685 version of Montaigne's *Essays* is still reckoned to be one of the classics.

The French writer in question was one of those types always on the edge of trouble, thanks to his vociferous skepticism regarding all forms of authority. At a time when all forms of authority still included people with power of thumbscrew, this was a risky game to play. Still, Montaigne (he wrote: "The only thing that's certain is that nothing is certain" on the ceiling beams of his study) managed to stay one step ahead of the theological sheriff. Traveling, on one occasion in 1581, to Florence, he was flabbergasted by the gardens of the Medici Villa Pratolino, where he saw all the latest hi-tech toys for boys (aka Renaissance princes). The gardens featured such amazements as moveable artificial scenery, water-powered organs and mechanical figures, and musical waterfalls. These marvels were the work of Bernardo Buontalenti, architect, engineer, and man for all seasons. Buontalenti was also extravagantly into theater (he set one up at the Uffizi for the Medicis in 1585), stage design, and boffo musical spectaculars of various types, whenever one of his princely patrons had something to celebrate. This, when he wasn't managing the Tuscany water supplies.

Speaking of which, the next drain on Medici finances was the ducal hiring of Sir Robert Dudley, a questionable English adventurer on the run (with his mistress) from a second wife and numerous daughters back home. Where, he claimed, if it hadn't been for family infighting and accusations of illegitimacy, he would have been earl of Warwick. In Italy he reclaimed marshland between Pisa and the sea, built canals, practically invented the city of Livorno, and introduced English shipbuilding techniques to the Medici navy. In 1647 he produced *The Secret of the Sea*, the first complete collection of charts based on Mercator's projection.

The hottest cartographers at the time were the Dutch, and in particular the Blaeu family. Father and son were mapmakers to the Dutch East India Company, whose main interests lay in exploring the fabled riches of the East and bringing back as much of them as they could lay their hands on. Blaeu *père* set up the family printing business in Amsterdam in 1602 and kept his maps up-to-date by nobbling captains straight off their returning ships so that he could monopolize the latest navigational poop.

Son Willem Jansoon Blaeu started his working life (and his interest in maritime matters) as clerk to a herring company. After which he became assistant to a Danish astronomer with a replacement metal nose, the irate Tycho Brahe. Who was walking home from his lab one evening in 1572 and saw the impossible—a new star, in heavens that were supposed at the time to be unchanging. The first guy to whom he pointed the nova out, and who didn't laugh at him for seeing things, was his friend the French ambassador, Charles de Dancey, venerable *bon viveur* and eminent good guy. One of whose recent charitable acts had been to go to Malmö prison and then offer to intercede, with the king, on behalf of an inmate. One of the period's eminent bad guys.

James Hepburn, fourth earl of Bothwell, was a swashbuckler for whom things had gone badly unbuckled. Fleeing Scotland for Scandinavia by ship, he had the extraordinary ill luck to fetch up in Bergen, where the local duke didn't believe Bothwell's tale of being a Scots gent looking for work, because this particular ruler had a female relative that Bothwell had done wrong by, years before. A habit of his. Fact being, Bothwell was on the lam because he had left his wife in the lurch back in Scotland, when it became clear he was about to be fingered for complicity in the murder of her previous husband. This would have been no big deal, had Mrs. Bothwell not been otherwise known as Mary Queen of Scots.

Who was in her own mess because she was a legitimate claimant to the English throne and this was not a good thing to be when that seat was already occupied by Queen Elizabeth. For a hothead conspirator like Mary it was inevitable that she'd end up losing hers. Elizabeth, meanwhile, had other things to worry about. Such as Sir Francis Drake et al. crossing the Atlantic and discovering to their horror that on the other side their compass needles no longer pointed at the North Star (where they thought it should). Crisis-level investigations conducted by Her Majesty's personal physician and science guru, William Gilbert, convinced him that the Earth was a giant magnet with a North Pole that attracted needles. This got a legion of other famed experimenters noodling away. One of whom, the mayor of Magdeburg, Germany, Otto von Guericke, fashioned an Earth model out of sulphur in 1650, rubbed it vigorously, and brought various needles close, to see which way they would point. (This was the heyday of propellor-heads, after all.) During one of his rubbing sessions, Guericke idly footnoted that the sulphur ball gave off a bang and a spark. Turned out to be electricity. Soon everybody was rubbing, including Francis Hauksbee.

In London in 1705, Hauksbee demonstrated an amazing electricity generator to the Royal Society. It consisted of a small, evacuated glass globe mounted on a spindle. When spun at high speed, and while a hand was held against the glass, a purple light appeared inside the globe. Nearby threads were attracted to stick to the glass. Hauksbee then moved on to finer things. He discovered that if two slim glass tubes were placed in a liquid, the liquid rose higher in the narrower tube. Since Hauksbee thought that this might be due to something happening between the liquid and the glass sides of the tube, he turned to the man whose middle name was "attraction": Isaac Newton. Newton explained capillary action in 1717.

By this time the great man was also the most famous person in the entire universe for clearing up the mystery of Kepler's "Holy Spirit Force" (remember?), and giving it the name by which both it and Newton are known today: gravity.

And that's enough of these weighty matters.

34

TICK TOCK

WALKING THROUGH PARLIAMENT Square in London the other day, I suddenly remembered that when I was a very small child during World War II, the sound of Big Ben ringing the hour on the radio was a comforting indication that Parliament hadn't been hit by incoming German V-1 missiles.

In 1859 the new Big Ben clock became the miracle of harrumph hi-tech precision (as had been promised by Prime Minister George Canning) when the gravity escapement system created by E. Beckett Denison was installed. Denison's trick for isolating the movement of the pendulum from that of the gears worked in such a way that no matter how much dirt or ice accumulated on the four sets of clock-hands, Big Ben would keep time accurate to the second. (Thank you, Fredericton, New Brunswick, on whose often-icy cathedral clock Denison had done a wet run.)

Now, your basic dead-beat escapement, on which Denison improved, links the movement of the pendulum to a couple of blades. As these move from side to side and turn about, each one catches on the clock drive-wheel and controls its turn on a shaft wound with cord attached to a weight. Perhaps the most famous early version of dead-beat escapement was the brainchild of George Graham, the greatest clockmaker of the eighteenth century. In the summer of 1736 one of his astronomically accurate clocks was taken to Sweden by a particularly arrogant French scientist, Pierre-Louis Moreau de Maupertuis, who was intent on proving his French colleagues wrong (and the English right) about the Earth's shape.

The fact that geodesy should have been at issue at all was not just another example of the Gallic chip on the shoulder that had been there ever since Newton. It was also the small matter of multiple shipwreck. Get the shape of the Earth wrong, and you also get wrong the measure of a degree across its surface. Which means that ships hit rocks that shouldn't be there, because the ships aren't where their navigators think they are. Entire fleets were blundering about in this fatal way, and far too often taking to the bottom large quantities of gold and guns. Maupertuis offered a solution: Using a one-degree difference in the position of a star (hence the need for precise timing), he calculated exactly how far across the planetary surface represented one degree of latitude. This took him into several months of mosquito hell in the middle of Swedish nowhere. And at the end of it, *sacre bleu*, Newton had been right! The Earth was an oblate sphere. No wonder Maupertuis made enemies back home in Paris. Pro-English, arrogant, and right.

Maupertuis became the target of savage attacks by Voltaire and was effectively hounded out of France. He later died at the home of Swiss mathematician pal, Johann Bernoulli, whose mathematical brother Jacob was deeply into why hanging chains assumed the position they did. This arcane catenary matter had originally been the obsession of a Dutch genius mathematician named Simon Stevin, who in 1585 also conceived of the novel idea of decimal coinage. No takers, until a one-legged American aristocrat, Gouverneur Morris, tossed it to Thomas Jefferson, who promptly made it his own. Hence dollars and cents. Undeterred by this, in 1794, Morris promoted another scheme (this eventually also snitched by another politican, DeWitt Clinton): to build a canal from Lake Erie to the Hudson River. The Erie Canal soon became a project famed in song and story, until somebody built a railroad across New York State in 1851, which pulled the plug on waterborne traffic.

By 1855 the New York and Erie Railroad had four thousand employees and logistical problems so complex that its general superintendent, Daniel McCallum, was obliged to invent modern business administration methods to deal with them: autonomous heads of division; professional middle management; daily, weekly, and monthly reports—all that stuff. And because one of the things regular and high-volume railroad deliveries of goods made possible was the high-turnover department store, places like Wanamaker's and Macy's were among the first businesses to adopt the railroad's own administrative pyramid structure.

Efficient delivery of an unprecedented range of items (from candelabra to tiaras to gloves) and production-line-inspired sales procedures left the new mass-market retailers with only one little problem: how to persuade the customer to cart purchases out the front door as fast as they were being hauled in the back. So the stores tried transforming their downtown emporia into something between a Victorian boudoir and the palace of Ramses II. With extras thrown in, including musicians, beauty salons, post offices, and nurseries. For the first time, shopping became a luxury experience. Once, that is, management had convinced its lady clients (the men wouldn't) to turn up and shop till they dropped.

Wanamaker's gave this crucial public relations exercise to N. W. Ayer, the first fully fledged advertising agency, and soon, near the end of the nineteenth century, a new breed of thinkers known as psychologists were being recruited to delve deeper into the consumer's emotional state to see if further profitable shopper manipulation might be on the cards. Harvard professor of physiology Walter B. Cannon (the only physiologist, I believe, ever to have had a mountain named after him) advanced these motivational matters further. With the use of the amazing new X-rays and a barium meal (which he invented), Cannon was able to study the peristaltic waves that accompany digestion and hunger, and he saw that these waves stopped suddenly if the subject were emotionally disturbed in any way. After years of experiment, in 1932 Cannon's masterwork, *The Wisdom of The Body*, introduced the idea of homeostasis: the maintenance of balance in the body's state by chemical feedback mechanisms.

Collaborating with Cannon was Mexican neurophysiologist Arturo Rosenbleuth, who in the early 1940s began to have conversations about feedback with a mathematical prodigy, the irascible Norbert Wiener of MIT. Wiener was interested in such phenomena as the way feedback works to ensure that when you pick up a glass of water and drink from it your mouth and glass manage to meet. He was concerned about such hits and misses because he was also working on the kind of equations that would facilitate antiaircraft artillery in potting their targets more frequently than once every twenty-five hundred shells (the average score on the south coast of England at the time—and no way to win a war).

The system Wiener came up with, named the M-9 Predictor, used feedback information provided by radar data on a target's track to help extrapolate where it would be, up in the sky, when the next shell arrived there. On that basis, his invention directed how the guns should be pointed. The gizmo worked so well that in the last week of large-scale German V-1 attacks on England, of 104 missiles launched against London only four made it through.

Which is why, thanks to Wiener, Big Ben chimed its way, undamaged, right through World War II.

High time I was out of here.

35 REBELLIOUS AFFAIRS

IN ONE OF London's oldest coffeehouses the other day I was sipping a cup (of Folger's, as it happens) and thinking about the first such watering-holes, like the one in St. Paul's Churchyard, where those who supported the American colonists' cause used to meet and greet in the 1760s. One of these dangerous coffee-drinking liberals was mathematician Richard Price, who put actuarial studies on a modern basis with his analysis of the Northampton Bills of Mortality, from which he was able to deduce life-and-death matters well enough for the new insurance companies to be able to charge realistic premiums and not go out of business before their clients did.

Price's statistical work attracted the attention of the French finance minister Anne-Robert-Jacques Turgot, who had the unenviable task of balancing the books at a time when the French economy was heading rapidly toiletward. One of Turgot's pals and advisers was the Marquis de Condorcet, whose claim to fame was the way he took the new statistical view of society to the next stage and invented what he called "social mathematics," with which he intended to predict all aspects of social behavior, and thus set the study of society onto a scientific footing.

Condorcet and Turgot both came out of the Physiocrat school of economic thought, which looked to the English example of a well-developed and free-trade agricultural market as the only way to save France from ruin. For them, the price of a loaf held the key to political stability. Well, as you know, they never made it out of the bakery in time to save their heads. But en route to Condorcet's eventual death in a French revolutionary prison, he tried hard to popularize the agronomic ideas of an English lawyer, Jethro Tull. In 1711 Tull had gone to the south of France for his health and had seen peasants in the vineyards around Frontignan hoeing the soil between their furrows and (although he didn't know it) in this way aerating the soil and making it easier for water to permeate. As a result of this, they were getting great crop yields without having to resort to the use of expensive manure (in those days, almost worth its weight in gold). When Tull got back to England and found that with this method he could grow corn in the same, unmanured field for thirteen years, he was in like Flynn with his farmer friends. His "how-to" book: *The New Horse-Hoeing Husbandry*, hit the shelves right in the middle of the great English Agricultural Revolution (crop rotation, new animal feed, fertilizers) and became an instant bestseller.

When Tull's ground-breaking work went into a second version, the editor was a journalist so famed for his prickly style he was known as "Peter Porcupine." William Cobbett had begun his writing career while teaching English to French immigrants in Philadelphia. Here, with remarkable disregard for the world around him, he wrote diatribes against upstart American ideas about democracy. In 1794 the eminent English scientist Joseph Priestley arrived, on the run from anti-American lab-burning mobs in England, and the pro-English Cobbett greeted him with an article lambasting the liberalism of both Priestley and his pro-American science pals back in England.

Priestley himself had started life as a minister of the church and at one time had run a Sunday School, where a member of his teaching staff was a guy named Rowland Hill, who would go on to renown when in later life he reformed the English mail service and introduced the first lick-it-and-stick-it, prepaid postage stamp: the Penny Black. Meanwhile, Hill's teaching career took off when he became teacher at the radical new Hazlewood School, in Birmingham. The place was gaslit and had central heating as well as a swimming pool, and the curriculum included such off-the-wall, New-Age stuff as applied math and modern languages. In 1822 Hill and his brother published a modest work titled *Public Education*, and this brought Hill to the attention of the left-leaning great and good, including one Robert Owen, enlightened mill owner and founder of what would one day become the British Socialist party.

Like many of his stripe before him, Owen spent time in America establishing one of the many (temporary) utopian communities fashionable at the time. His was in New Harmony, Indiana, and when the commune failed and Owen returned to England, his sons remained in the United States and became citizens. One of them, Robert Dale Owen, went on to become a leading figure in Indiana politics, spending two terms in the U.S. House of Representatives (where he introduced the bill to set up the Smithsonian). In 1888, his daughter Rosamund married (for one week, whereupon the new husband inconsiderately died) fervent Zionist supporter and spiritualist Laurence Oliphant. Earlier in life, Oliphant had established himself as travel writer for the *Times* of London, covered the Crimean War, and worked as a British spy. At one point he also took a job as secretary to Lord Elgin, who was governor-general of Canada but who went down in history as the son of the man who, in 1803, snitched the Elgin Marbles. Or, as Elgin *pater* would have put it: "removed them for their own good." The marbles (which subsequently ended up in the British Museum) consisted principally of large bits of the frieze of the Acropolis, dating from the fifth century B.C. And the Acropolis, at the time, belonged to the Turkish occupying power, which couldn't have cared less about a building that was in any case gradually being pulverized for lime.

It took thirteen years to transport the marbles back to England and sell them to the British Government for a price that ruined the Elgin family fortunes for two generations. But since Elgin was in dire straits at the time, he was in no position to haggle. One of those promoting the idea of the marble acquisition was Sir Thomas Lawrence, painter to the king and general artistic big cheese. He'd started life as a child prodigy, and by this time was charging such eminences as the duke of Wellington and the Prince Regent an arm and a leg to do their likeness. Lawrence was already so famous that back in 1792, when he had become president of the Royal Academy, he had also been elected painter to the Dilettanti Society, entry to which required that you (a) were aristo and (b) had crossed the Alps in search of culture. Lawrence wasn't, and hadn't, but he was a friend of Sir Joseph Banks.

Since Banks was president of the Royal Society at the time, and chummy with the king, the Dilettanti rules were waived for Lawrence (you get the feel for the "if you've got it, flaunt it" flavor of the times, right?). But Banks *did* have what his contemporaries took to be an impeccable scientific background—that is, born rich with good family connections and an obsession for botany. So, early on, it wasn't hard for him to win the coveted post of naturalist on Captain James Cook's first voyage of exploration to the Pacific in 1768. Their various landfalls inspired Banks later to promote the idea of Australia as the absolutely perfect spot to dump convicted criminals, and to organize another kind of transplantation: that of Tahiti breadfruit trees to the West Indies.

The first ship sent on this mission was the *Bounty*, of mutiny fame. Nothing further was heard of the mutineers until their descendants were discovered thirty years later, on Pitcairn Island, where the local scuttlebutt held that their leader, Fletcher Christian, had secretly made it back to England and a life of obscurity, thanks to the obliging captain of a passing American ship.

The captain's name was Folger.

36

LOCAL COLOR

WE'VE BEEN DOING some home decorating recently and I was rejecting a particularly bilious shade of avocado when I suddenly remembered that the French empress Eugénie (wife of Emperor Napoleon III) went to the Paris Opera one night in 1863 and blew everybody away because she was wearing a green silk dress.

Not just any old green, though. This was Malachite green. The latest hi-tech product to come out of that well-known chemical treasure-trove of which I have spoken many times, coal tar. Earlier that year a German chemist called Lucius had discovered the color. Apart from astonishing operagoers, the new dye also contributed to the success of Lucius's company (which later changed its name to Hoechst). And the reason Eugénie's dress was such a sensation was that the green didn't look blue under gaslight.

Now, I don't know what opera Eugénie heard that night. Pity she never heard *Carmen*, which would not be performed until 1875, after she had already gone into exile. That particular opera story line had been provided by Eugénie's mother, the countess of Montijo, in Spain. Back in 1830 she had told the story to Prosper Mérimée, a traveling art student who was later to become a writer, when the count had invited him back to the Montijo family home, and Mérimée then charmed everybody, including the five-year-old Eugenia (as she was then called). Mérimée went on to snitch the countess's tale of *Carmen* for his novel of the same name, which Bizet would later snitch for his opera of the same name.

Mérimée remained in close touch with Eugenia. This may be why, years later when she met and married the emperor, she mentioned her best pal Prosper, who was promptly given the title of senator and oodles of boodle to support him while he wrote plays and novels. Just as well the emperor didn't know Prosper had been having an affair (one of many) with his uncle's mistress. Sexual peccadilloes were a habit Prosper failed to break for most of his life, starting with a scandal while he was still at school.

Where one of his best friends was a far less interesting

type named Adrien de Jussieu. The last in a long line of botanists. Who took over from his old man when the latter retired from being professor at the Paris Natural History Museum. Followed in the footsteps, really, and not much else. But his daughter married a guy who really *did* make a mark. Name of Armand Fizeau.

In 1849 he worked out the speed of light with a crafty gizmo consisting of a 720-tooth cogwheel, spinning very fast. Fizeau shone a light beam through the cogs, at a mirror, several kilometers away behind the cogwheel, which reflected the light when there wasn't a cog in the way. By relating the speed of the wheel to the point at which the cogs eclipsed the light, Fizeau was able to say that light moved at 315,000 kilometers per second. Pretty close, for 1849.

His close collaborator in all this was, like Fizeau, an ex–medical student who couldn't stand the sight of blood—Léon Foucault. In 1845 both of them took the first clear daguerreotype photographs of the sun's surface. It was working out a mechanism for keeping the camera pointed at the sun (and later, stars) for long exposure times that led Foucault to invent his great pendulum (see another essay for a swinging description). That same year, Foucault had met a diffident young Scotsman, William Thomson, while studying at Henri Regnaut's lab in Paris. Thomson (later Lord Kelvin of absolute zero fame, in one sense) went on to similarly great things. Least of which was his theory that explained the way certain crystals responded to changes in temperature by becoming magnetic. He showed that there was a relationship between temperature and the permanent polarity of these crystals.

Back in 1824 this phenomenon had been named "pyroelectricity" by Sir David Brewster. This hardy Scot had failed to be a tutor, magazine editor, love poet, and preacher, so he settled for science, concentrating on polarization of various kinds and inventing the kaleidoscope. The instrument was primarily to be used for the designing of new patterns in carpets, wallpapers, and fabrics. Well, why not? Eventually, through what reference books describe as being "zealous in advancing long-neglected areas of science" (he sounds like a modern Ph.D. candidate), Brewster proceeded by leaps and bounds to important medals and the principalship of Edinburgh University.

Brewster married Juliet, youngest daughter of James MacPherson, the man who changed the path of cultural history when, after touring the Scottish Highlands, he pretended to have discovered a great third-century Gaelic epic poem written by a Celt called Ossian. This apparently archetypal bit of ancient European self-expression hit the philosophical world (in particular the German) like an earthquake. With its portrayal of an earlier, simpler existence, the epic almost single-handedly triggered the Romantic movement. Its portrayal of an ancient warrior society of super-beings would one day give the Nazis a few ideas. And Thomas Jefferson thought Ossian must have been "the greatest poet that ever existed!" Not bad for a fake. Two years after the fake epic was published and MacPherson had become a literary lion in London society, in 1763 (thanks to the good offices of the earl of Bute) he was offered the job of secretary to the governor of Florida and left for America.

Bute himself is remembered as a thoroughly unpleasant politico who was prime minister for a year. And is forgotten for his pivotal influence in getting the Prince of Wales to set up what was to become one of the greatest botanic gardens in the world at Kew, outside London. Bute had been a keen gardener for years and in 1757 he persuaded the prince to appoint Sir William Chambers as architect to the Gardens. Four years later Chambers built the amazing Kew Pagoda, 163 feet of the most detailed example of *chinoiserie* in Europe, and quite something if you like that kind of stuff. Conforming to the overblown style of the times, he also added a mosque, an "Alhambra," various classical temples, and an imitation Gothic cathedral. This made Chambers the master of the garden temple genre, and a cert for the commission to build Somerset House, a gloomy, grandiose pile that was until recently the home of the gloomy and grandiose British Internal Revenue Service.

Chambers shared the commission for Somerset House with Robert Adam, one of the most influential stylists in architectural history. This was the man who, when he got back home from a Grand Tour of Europe discovering the beauties of Roman and Greek ruins, turned façades into an art form. Adam set about persuading those among the English aristocracy with more money than sense that their moldering piles would look much better with Doric front doors and colonnaded wings. When he'd finished, anybody who was anybody lived in something that looked like a bank.

Adam's runaway success meant he had many imitators. One of whom, George Dance, "Adamized" a minor stately home named Camden Place, in the Kentish village of Chislehurst, south of London. At one point, late in the nineteenth century, it was rented by its minor aristocrat owner to an exiled French empress who wore green silk dresses.

And that's enough local color for now.

37

DOES THIS TAKE YOU BACK?

"DISEASE: NOSTALGIA. SYMPTOM: *an irrepressible desire to go home.*" That plaintive entry caught my eye recently while I was leafing through a dusty volume on quackery that was otherwise so dull it would cure your insomnia. The line occurred in the section devoted to nosology, the classification of disease by symptoms, a rave craze in the eighteenth-century medical world.

Turns out, the top nosology know-all of the late 1760s lived in Scotland. Name of William Cullen, he was professor of the theory of medicine at the recently opened Edinburgh University. And author, throughout a long and distinguished career, of one single, underwhelming research paper on evaporating fluids. (Back then, you didn't perish if you didn't publish.)

William Cullen's star pupil (and the guy who would succeed him as prof., in 1766) was Joseph Black, famous for at least three things. He always carried a green silk umbrella, he discovered latent heat (and so was able to tell James Watt how to make his steam engine work), and he founded a dining fraternity in Edinburgh known as the Oyster Club. This select nosherie was the regular elite-meet for most of the luminaries of the Scottish Renaissance (*Dictionary of Scientific Biography*'s phrase, not mine). Among those to be heard enlightening the world over the seafood each week were economist Adam Smith, et al. Al included a today-half-forgotten expert on the circulatory system (and good pal of Joseph Black's), James Hutton.

I suppose Hutton was a great example of the Scottish Renaissance Man the *DSB* had in mind. He studied humanities, physics, geography, law, medicine, and chemistry. Then qualified as a doctor. Then, in the manner of such people, became a farmer. Well, why not? It may have been a consequent landowner's interest in rocks and soil that got him into geology. From 1764 he began a series of trips to various of the stonier parts of the British Isles, tapping and chipping away. The chief object of his attentions tended to be basalt, since of the sundry explanations available at the time regarding the formation of the Earth, Hutton was most attached to the molten-interior, liquid-granite hypothesis.

Well, all the hammering must have been really productive, because in 1785 Hutton penned the outline of a modest work, eventually to be published under the not-so-modest title *Theory of the Earth*, and blew everybody away with his description of a great, cyclical process—degradation of land by erosion, the resultant deposits washing into the sea and settling over millions of years to become sedimentary layers, eventually to be thrown up again, to be eroded once more, and so on. And, said Hutton, if this process had taken as long in the past as it seemed to take in the modern world, then the planet was humungously ancient, never mind what the Bible said about six days. It would be this particular bit of Hutton's geological uniformitarianism (the fancy name for his theory) that would eventually inspire Darwin.

Hutton was accompanied on some of his peregrinations by another Edinburgh pal, George Clerk, also an Oyster Club member and amateur rockhound. Clerk goes down in history for his *Essay on Naval Tactics*, reputed to have inspired Admiral Horatio Nelson to victory on the *Victory*. Clerk's other notable feat was to marry the sister of Robert Adam, one of the hottest British architects of the eighteenth century. You wanted a scrambled hash of oh-so-chic Neoclassical bits and pieces all over your stately home interiors, you hired Robert, who would transform your crumbling pile into instant pseudo-Graeco-Roman for only a fortune. One of Adam's more fastidious imitators was a cabinetmaker and furniture freak named George Hepplewhite, who added simplicity and elegance to some of Adam's more extravagant chairs, and in 1788 came out with his own bestselling *Cabinet Maker and Upholsterer's*

Guide, whose designs were to be often copied (in America) but rarely acknowledged (in England).

Hepplewhite's guide included instructions for japanning mahogany. Japan lacquer (so-called because in the seventeenth century, when lacquer arrived in the West from China, nobody knew the difference between one place and the other) became so popular among European royals that you needed a King's ransom to buy some. The Chinese wouldn't reveal the secret of lacquer manufacture, so buyers took a real shellacking. Till 1732, when Thomas Allgood, in the Welsh town of Pontypool, came up with a new kind of lacquer that became known as known as "Pontypool Japan." If you possess one of your great-great-grandmother's tea-caddies with Chinese figures on it, you will know what I mean. The great advantage of Allgood's mix of linseed oil, umber, litharge, pitch, and turpentine was that it was cheaper and more available than the real thing. It could also be put onto tin, which was cheaper and more available than wood. If, that is, you knew how to make tin. For which you had to know how to roll sheets of iron.

Not surprisingly, one of the best iron-rolling mills around at the time was, guess where, in Pontypool. At one point, one of the Allgood family worked at the mill for its owner, John Hanbury, who had developed a technique for processing red-hot iron through several sets of rollers to make it extremely thin, whereupon it could be dipped in molten tin and shaped into utensils ready for Pontypool japanning. When Hanbury originally set up his mill, the cutting edge in sliced sheet-iron manufacture was at Stjarnsund, Sweden, where an unsung genius named Christopher Polhem designed and built amazing water-powered machines that would do anything you wanted to hot metal. Polhem had started out as a mining engineer, and his work would end up giving Sweden the reputation for expertise in metallurgy it still has today. Not much is known about him because he got biographied by a young admirer whose reputation then totally eclipsed that of Polhem. This, because after the two of them had worked together on the Swedish Board of Mines, the guy in question went on to what I suppose might be described as higher things.

Emmanuel Swedenborg (you're probably there before me) was yet *another* polymath. He did humanities, geology, metallurgy, paleontology, flying machines, submarines, started the first Swedish science journal, and dabbled in astronomy. It may have been the early stones-and-bones work that got him looking into the origin-of-life stuff, but when he started thinking about the soul (and postulated its seat in the cellular cortex), he was ready for bigger things and planned a great work on nothing less than Creation. But . . . the best-laid plans. In 1745 he had an epiphanal vision, in which God ordered him to dump science and technology in favor of the Bible. Doing so changed Swedenborg's life from that of propellor-head to that of prophet of his Church of the New Jerusalem.

One of the major Swedenborgian American commentators was a journalist and businessman named John Bigelow, who, in 1849, became managing editor and part owner of the *New York Evening Post.* His co-owner at the *Post* was one of America's best-known Romantic poets, name of Bryant, who wrote about the woods and streams of his native Berkshires. All his life he said he wanted to return there. Not surprisingly, for such a nostalgic type, his first names were William Cullen.

Does that take you back (to the beginning of this essay)?

38
OOPS

FROM TIME TO time it gives me great pleasure to come across, and publicize, the name of somebody who never got the credit. In this case, peering through my optically pure reading glasses in the British Library, there was the name of somebody who, in the privacy of his later years, must at least have reacted to events with an "oops."

Name of Chester Moor Hall. Who?? OK, try this one: John Dollond. Right! The guy who invented the optically pure achromatic lens in 1747. Nope. Hall does it, years before. Nice guy, Hall. And like many such, finishes last. Around 1729 he becomes convinced that it's possible to make a lens that will not give you all kinds of blurred, colored images. This is the bane of astronomers up to then, and why they are seeing things like "planets with ears" (Saturn). Around 1729 Hall sticks lenses of differing densities (flint and crown glass) together, and the different dispersions kind of cancel each other out. Bingo. Achromatic. No-color-fringes. Total clarity. So Hall makes a couple of telescopes for friends, sticks the whole experiment in the closet, and goes back to being a land-owning magistrate in Essex, England. Never does anything about it. Even when he hears about some other guy (name of Champness) fussing about having invented it before Dollond. It takes another hundred years for the news of Chester Moor Hall's work to get mentioned in a paper to the Royal Society (where it's then filed and forgotten). And then another hundred and sixty-five, to get to me. Forefront of historical research I'm *not*. But you knew that.

Meanwhile, John Dollond. Gets out of silk-weaving (the d'Holland family had originally been Dutch textile manufacturers) into optics and sets up in business with his son, in London. Instrument makers to the gentry. Makes a fortune. Dies leaving the rights to his patents (for many gizmos including the achromatic lens) split among his heirs, one of whom is the husband of Dollond's no-fool-marry-the-boss's-daughter-Sarah apprentice, Jesse Ramsden. For Ramsden, this is a marriage made in heaven. He has his hands on the lens, plus he's a real whiz with metal. And particularly good at making marks on it. Degrees and minutes of arc, that stuff. Which he proceeds to do, on a thousand sextants he then makes for the British Navy and various explorers. The thing about Ramsden's marks is, they're extremely accurate. And the more accurate the marks on your instrument, the more accurate your measure of whatever it is you're measuring. In the case of navigators, using one of Ramsden's things means you're more likely to miss the rocks.

As long, that is, as you know the rocks are there in the first place. Not widespread, that kind of knowledge, back when the Industrial Revolution is beginning to require the import of thousands of tons of near-freebie raw materials from your new colonies, and the export of finished goods to the same. All the Brits have to do, to facilitate this sweetheart import-export deal, is pass colonial laws to oblige the hapless colonists to sell to, and buy from, *only* Britain. And do it all in *only* British ships. Which annoys certain colonists enough, the result is the U.S.A. Still, the scam works long enough elsewhere (India, Africa) to pay for most of the culture and the stately homes and the advanced nation status that the modern tourist now enjoys on a visit to Britain.

All those profits are why, back then, lots of heavily laden, low-in-the-water ships are racing back and forth, and all too often getting more low-in-the-water than you might wish (after hitting the aforementioned rocks). Hence the great lighthouse-building mania of the period. Now, putting up lighthouses has been no big deal since the Pharaohs. But the problem about the late-eighteenth-century lighthouse isn't the "house" bit. It's the "light" bit. As in: candles nobody can see, until it's too late. Since this is a bottom-line profit-related matter, things get done about it. Thanks to Jesse Ramsden's accurate marks. In this case, on a giant four-foot theodolite that makes possible the Ordnance Survey of Ireland. Once the surveying teams can see what it is they're pointing the theodolite at. Not easy, early on, in the murk of Ulster. Until (1824) enter a young army type, Thomas Drummond, who invents one of those gizmos that go into the language. Works so: Jet of hydrogen and oxygen, it burns, is directed at small ball of lime, it goes incandescent, this is reflected by parabolic mirror. Limelight. Great for surveyors. And actors. And mariners heading for the rocks. Unless the lighthouses run out of gas. They do.

So in 1849, Belgian prof. Floris Nollet makes all the gas you might ever want, with electrolysis. Positive and negative electrodes in water. Current between them disassociates hydrogen and oxygen from the water. Problem solved. Unless the lighthouses run out of electricity. They do. Advances are made (remember, this is a bottom-line matter), and in 1871 a Nollet worker, Zénobie Gramme, invents the dynamo. Wire coils spin in a magnetic field and produce electricity. Enough to power an arc light. Two carbon rods almost touch. Electricity coming through the rods jumps the gap between them, sparks, and makes the carbon points go incandescent. Makes limelight look dark. And makes possible the electric arc furnace. In which you can use two carbon electrodes to create unprecedentedly hot heat.

Enough for a Frenchman in 1892 to think he can make artificial diamonds by heating iron and carbonized sugar hot enough to dissolve the carbon into the iron. He will then rapidly cool the iron in water, so it solidifies with enormous pressure. This will cause microscopically small carbon particles (diamonds) to appear. The furnace inventor and would-be diamond merchant, Henri Moissan, thinks what he comes up with are the aforementioned synthetic stones. Well, as it turns out, they aren't.

But nobody cares, since the arc furnace has dazzled the scientific community far more than Moissan's little carbon chips. Because the furnace is an amazing new fun tool to do any amazing new fun things people can think up in their (now-) high-temperature chemistry labs. They do. In 1895, Moissan puts a mix of lime and carbon in the furnace, heats it to two thousand degrees Centigrade, and gets some stuff that is pretty dull until he brings it into contact with water, when it gives off a gas. Acetylene. Which is pretty dull until he sets light to it. Makes arc light look dark. By 1899 there are acetylene plants up the ying-yang, most of them in places like Niagara Falls, the Pyrenees, Norway, and Switzerland, where falling water generates the electricity required for the furnaces. Until Edison. When *he* does what he does, the bottom drops out of the acetylene light market.

So by 1912, lonely acetylene freaks are bumbling around looking for anything they might be able to do with their stuff. In Stuttgart, one of the chemists at Greisheim Electron, on the lookout for some kind of doping material with which to weatherproof the fabric on aircraft wings, tries a mix of acetylene, hydrogen chloride, and mercury. No good. So he sets the stuff aside on a sunny windowsill and later notices that it forms a milky sludge, which then goes solid. He files a patent and forgets it. And in 1925 the patent lapses. Which is why I end, as I began, with a guy whose favorite word must have been "oops."

He was named Fritz Klatte. The sludge was named PVC. The first plastic.

39
TEA, ANYONE?

I WAS RECENTLY doing some reading with one hand while with the other abstractedly stirring some sugar (too much, it turned out, when I made the mistake of drinking some) into my cuppa, thinking about how the English tea-drinking thing is all a myth. It was the Dutch who really started the craze.

In 1610 the first shipment of *chai* from China arrived in Amsterdam and turned Holland into a nation of addicts. Within a few years the eminent Dutch physician Cornelius Bontekoe was prescribing two hundred cups a day for the general health. By 1650 the Dutch East India Company was importing thousands of tons of tea, re-exporting it as far as New York, making a million, and going back east for more. Tea (and the porcelain cups it went into) made Holland very rich. And paid for all those instrument-makers working on things like barometers and telescopes and such, that would make it even easier for Company navigators to find China every time, pick up the magic leaves, and then head for home. And find Amsterdam every time.

As usual, this was another case of bank account drives science and technology. Profitable precision at sea required instruments, and in turn these soon made possible precision in measurement of all kinds. Which is what drove Dutchman Gabriel Fahrenheit to do what he did, around 1713. Well, not exactly. Research reveals (as it does so often) that the accepted view is wrong. In 1708 Fahrenheit snitched the thermometer for which he is famous from the ex-mayor of Copenhagen. This guy, Ole Romer, had come up with the idea of a heat scale way ahead of Fahrenheit. All the latter did was to fiddle with the numbers on the instrument. Shortly thereafter, by fortunate happenstance, all Romer's research notes were destroyed in a fire, and Fahrenheit was home free.

Still, Romer had already made his name for other, more cosmic reasons. Earlier, in 1671, he had been picked up by a passing French astronomer, Jean Picard, who was on his way to identify the exact position of Tycho Brahe's astronomy center at Uraniborg (on the island of Hven, where the latter had previously taken certain crucial stellar fixes), so as to get some geodetic matter or other sorted out. Picard persuaded Romer to move back to Paris with him, to work in the posh new observatory there. Romer's subsequent discovery of the differing times of the eclipses of Jupiter's moon Io convinced him that these related to the differing Earth-Jupiter distances, and concluded from this that the speed of light was finite and not instantaneous as had been thought since Aristotle. Big deal, this. Romer is all too unfairly ignored in modern classrooms.

His pal Picard was a kind of scientific general factotum who did science things for Louis XIV. Louis, being a divine-right monarch and vested with power of thumbscrew, tended to get what he wanted. At this point (1674) what he wanted was water. Problem being the fountains and pools and water-powered amusements at his new palace of Versailles. Which weren't doing what they were supposed to, because for some reason the water supply wouldn't supply. To Picard, who had recently worked out the degree of Earth's meridian to within a few feet, this little difficulty was a mere bagatelle, and in no time at all his telescope level helped identify the discrepancy causing all the problems. This turned out to be pretty obvious: the palace at Versailles was very slightly higher than the surrounding terrain and source of water. Adjustments to cisterns, reservoirs, and channels were made, and water soon flowed freely.

Just as well, since the other little thing the king had in mind for Versailles was the biggest garden this side of Babylon. Visit the place. You'll see why Le Notre, the king's horticulturalist, must have thought Nature needed a haircut. His Disneyfication of a large bit of the French countryside took him and his thirty-six thousand laborers over twenty years, became the salon topic of choice for the chattering classes, and made it fashionable for French aristos to get dirt under their fingernails. One of the new genteel grubbers was a fellow named Duhamel du Monceau, who set up one of the first arboreta on his estate at Château Denainvilliers and wrote about manure and muck and such. And, since this led him to intimate knowledge of everything that happened between sapling and great oak, naturally enough Duhamel moved on to become inspector-general of Marine. Not such an odd career move, when you consider that this was the time when it took a thousand oaks to make a warship. As a result of which French forests were soon reserved for nothing but.

As a young sprig back in the 1720s Duhamel had picked up this love of matters botanical during lectures he attended at the Jardin du Roi, where he also became great pals with one Bernard de Jussieu, who was running the Garden's field trips at the time and winding up to yet another of the many and varied plant classification systems glutting the libraries. Jussieu came from three generations of classifying botanists, the last of which, Adrien (son of Bernard's nephew), did his bit for the family with a piece excitingly titled *Vegetable Taxonomy.* Adrien also set up a herbarium at the French Natural History Museum, together with yet another scion of yet another down-to-earth family, Adolphe-Théodore Brongniart.

Adolphe was a perfect example of what to do when every inch of your preferred research field is already crowded out with other researchers. He dug deep, and came up with something that would become virtually his own. Paleobotany (in Adolphe-Théodore's case, the morphology of fossil plants) was a subject virtually untouched, except for some earlier work by a minor Scotsman whose publications were pretty thin on the ground. Around 1815, the minor Scot in question, William Nicol (lecturer, natural philosophy, Edinburgh University), had used Canada balsam to cement pieces of fossil wood or minerals onto a glass plate and then ground the sample down to slices so fine you could see through them with a microscope and discover all kinds of good stuff (like bubbles in crystals, that told you something of the way the minerals had been formed, or cell patterns showing what kind of plant the sample came from).

In 1828 Nicol stuck two bits of an Iceland spar crystal together (with Canada balsam), and invented the Nicol prism. Iceland spar splits a beam of light into two polarized rays (a fact as it happens, discovered by Ole Romer's father-in-law, Erasmus Bartholin). If two Nicol prisms were used together, when the second prism was rotated, one of the polarized light rays coming through would dim and then disappear, once it had rotated through ninety degrees. Fascinating, but not everybody's cup of tea, right? Wrong. In the 1830s a Frenchman, Jean-Baptiste Biot, discovered that some liquids would twist the polarity of the ray, and the degree of twist would depend on the type and concentration of the material in solution. The twist in polarity was of course easily measured by a Nicol prism.

In 1845 a Parisian optician, Jean-Baptiste-Francois Soleil, perfected an instrument that would do all this and revolutionized life for the beverage drinker. Soleil's new gizmo was known as a saccharimeter.

With it, I could have told in advance that the tea I was drinking at the start of this essay would be too sweet.

40

A LIGHT LITTLE NUMBER

I POURED MYSELF a glass of delicious Bordeaux recently before settling down to watch one of the more spectacular things you see on the box these days, and as I idly sipped, my eye fell on the wine-bottle label and the number on it. So I knew it was going to be a pleasant little vintage, though not too big. The percentage-alcohol-by-volume-number was only 11.

Meanwhile, the Shuttle liftoff happened and I was, as ever, glued to my seat. I'm gripped by everything to do with the Shuttle, but in particular the delicate way the pilot gets to position the seventy-eight-ton vehicle to within half a degree so he can do orbital delivery runs, or pick-ups and drop-offs as required. These delivery stops are accomplished with the aid of forty-four tiny jets all round the Shuttle, some of which can produce as little as twenty-five pounds of thrust, thanks to more-bang-for-your-buck hypergolic fuel, one part of which is stuff called hydrazine.

Apart from these celestial uses, hydrazine features in more humdrum forms such as rust control in hot-water systems, pharmaceuticals, photography, and plastics. And something related to my glass of French red: hydrazine is also a fungicide. The first of which appeared in Bordeaux. Suicide capital of the world in the 1880s. If, that is, you had been a winemaker, and now weren't, thanks to the devastating effects of downy mildew. This little killer fungus arrived on vine stock brought in from America to replace the vines earlier destroyed by *phylloxera*, itself arriving on American stock brought in to fix an earlier plague. When downy mildew first appeared in 1878 it was jump-out-the-window time for anybody left in the business. Until 1882, when Prof. Pierre Marie Alexis Millardet of Bordeaux University came up with a fungicide mix of lime, copper sulphate and water, and the winemakers' troubles were over.

Millardet learned all he knew from Anton de Bary, his teacher at Strasbourg U., who goes in the history books under "Father of Mushrooms" (well: "mycology") because he was. Up to De Bary, people thought fungi were products of the plant they grew on. De Bary showed they were symbiotes (he coined the word). So, as you spray today, you know whom to thank.

De Bary himself had started in medicine, in Berlin, under the great physiologist Johann Müller (author of the classic *Handbook of Physiology*, 1840). This was the guy who finally knocked on the head the old nature-philosophy speculative guff that permeated medicine at the time and that involved everything from supernatural cures to animal magnetism and negative forces. Müller had his own bouts with the negative, at one point becoming *so* suicidally depressed he went to Ostende, Belgium. In 1847 while he was Berlin dean, he appointed one of his brighter students, Rudolph Virchow, to the post of instructor.

A year later everything hit the fan in Germany with the revolution of 1848 and Virchow was fighting on the barricades. Social issues then became the bedrock of all of Virchow's work, fundamental to which was his discovery of the basic function of the cell. Virchow saw the human body like a democratic society, a free state of equal individuals, a federation of cells. And all disease was nothing more than a change in the condition of the cells. These egalitarian views led Virchow to help set up the German Progressive party in 1861, and four years later provoked none other than the powerful Otto von Bismarck to challenge him to a duel. Fortunately for Virchow, nothing came of it, and Virchow went on to become so harrumph and wield so much power himself he was eventually known as the "pope of German medicine."

During a short teaching break at Wurzburg, Virchow taught a certain Victor Hensen, who made his mark with studies of the hearing organs in grashoppers' forelegs and the identification of a couple of bits of the human cochlea. In 1889 Hensen sailed all over the Atlantic for 115 days in search of another of his obsessions. For which task he had designed a special net made of uniformly woven silk, normally used by millers to separate different grades of flour. Henson's target was invisible, tiny, and everywhere. Well, everywhere there were floating nutrients. And when they had eaten up these nutrients, Hensen's obsession, plankton, would die and sink to the bottom of the ocean, where over a zillion years their shells would form sedimentary rocks.

By the early twentieth century these rocks were being ground up into a fine powder called *kieselguhr*. One use for which was most of the stick in a stick of dynamite (*kieselguhr* being uniquely inert, a valuable property when being used to blot up nitroglycerine). Another *kieselguhr* use was more mundane. When tiny amounts of nickel were deposited onto *kieselguhr*, the nickel acted like a catalyst to encourage hydrogen molecules to combine with oil molecules and make oil hard enough at room temperature to spread on bread (when the oil was palm oil, and part of margarine, that is).

I've mentioned the 1869 inventor of margarine before. Frenchman name of Meges-Mourriès, who also patented an idea for effervescent tablets and (in 1845) the use of egg-yolks in tanning leather. These were great days for nerds. Meges-Mourriès's margarine work (and his Legion d'Honneur medal) was suggested by France's Great Man of Chemistry, Michel-Eugène Chevreul, who wrote the book on fats during his nearly ninety years at the Paris Museum of Natural History. Chevreul discovered and analyzed every fatty acid, named them all, and turned the haphazard business of the soap boiler into an exact science. With what Chevreul was able to tell them, manufacturers could now make soap cheaper and better. Chevreul's fatty knowledge also made possible brighter candles. Small wonder that, when the man who thus made the world clean and light died aged 102, the entire country took the day off in mourning.

One of the products needed for soap was alkali, and when Chevreul had begun his work, this was still provided from wood ash. But as most of the French forests began to go down to the axes of the naval shipbuilders, alternative sources were being avidly sought both by the French and by their opponents at the time, the British. One such alternative turned out to lie along the rocky coasts of Britanny and Western Scotland. It was a seaweed known as kelp. Turning it into ash was simple and profitable. Your peasants raked it off the rocks and burned it in a pit with stones pressed on it so that it made a large, hard cake, which could be ground up for later use. In Scotland the kelp ash replacement financed many a new pseudo-Gothic castle for the owners of the previously valueless rocky beaches.

In 1811 France the kelp ash went into the nitre beds of Bernard Courtois, gunpowder manufacturer and accidental discoverer of a new element. That year Courtois leached his ash with water, ready to evaporate the salts he needed. When he added an overgenerous dollop of sulphuric acid (to get rid of the unwanted sulphur compounds) he was engulfed in violet fumes from the vat. Further investigation revealed violet crystals. Two years later the violet bits had been analyzed by one of France's foremost analyzers and given the name, after the Greek for their color, iodine.

The analyst in question, Joseph-Louis Gay-Lussac, also did the contemporary equivalent of the gripping stuff I was watching on TV at the start of this essay. He went up to a record height (in a balloon) to do science. It was also Gay-Lussac who told me my launch-watching wine would be light.

That alcohol-by-volume figure on the label (remember?) is known as the "Gay-Lussac number."

41 LEND ME YOUR EAR

I WAS COCKING an ear to the (usually unreliable) weather forecast on radio last night and remembered one of history's unsung heroes. So let's hear it for William Ferrel, a shy, self-taught schoolteacher from Pennsylvania who, in 1858, first explained the way the Earth's rotation affects how the weather moves and came up with the math to fit. I say "unsung" because most people think it was all done by a French guy named Coriolis. *Non.*

Ferrel, like everybody in nineteenth-century weather work (and indeed those involved in almost anything to do with science), was inspired by the work of Alexander von Humboldt. It's difficult to say what Humboldt *didn't* do. Just a taste of what he *did* do: economics, geology, mining, electricity, climatology, geography, oceanography, cosmology, math, exploration (everywhere), vulcanology, botany, chemistry, surveying, and isn't that enough? Humboldt was the maven's maven. He was also one of the first true ecologists. One of the other f.t.e.'s was the guy he visited in 1804, after a strenuous trip around South America, Thomas Jefferson. Who was at this time deeply into Lewis and Clark expeditions, coastal surveys, agricultural improvements, and such. So he and Humboldt got on like a house on fire.

They also shared a background deeply influenced by philosophy. Humboldt by Kant, Jefferson by the faculty at the College of William and Mary, second-oldest university in the United States, where he'd studied the subject before going on to other things like being U.S. prez. The College of William and Mary may not entirely appreciate this remark, but I read somewhere that a lot of their foundation money came from one Lionel Wafer, surgeon, writer, and buccaneer. Who flaunted body tattoos and a lip plate (courtesy of the Darien Indians of Panama). And lots of loot, ill-gotten over five years of piracy. Well, all good things . . . so when he got caught and went to the pokey in Jamestown for two years, the aforesaid loot was confiscated, to be "applied to the building of a college."

Interestingly enough, pirate loot wasn't always gold bullion and jewels and such. Sometimes the Spanish galleons that got knocked off were carrying *really* valuable cargoes, like crushed, unimpregnated female beetles of the cochineal family. Brushed off Mexican cacti early in the season, stove-dried, and ground, by the 1620s these little thingies were in the hands of a Dutch noodler named Cornelius Drebbel, who was about to change life for the military. Drebbel made microscopes, submarines, magic lanterns, and scarlet dye. This last involved an accident he had, some time after moving to London, with cochineal beetles. Which he let fall into a mix of sulphuric and hydrochloric acid in a tin mug. And noticed the amazingly scarlet result. Told his son-in-law, Abraham Kuffler, who promptly produced a brand-new dye named Kuffler Scarlet. In 1645 Cromwell dyed the uniforms of his new Model Army with it, and from then on, the term for anything British, marching, and carrying a weapon was "Redcoat."

If in the mouth of a Scot, Redcoat was generally a term of opprobrium, given the way the British behaved after the failure of the 1745 Highland uprising, led by Bonnie Prince Charlie. As a result of which life in Scotland became so unpleasant the Scots fled to America in droves. Many years later, when the dust had settled, one or two fugitives went inconspicuously back home. As did Flora MacDonald (who'd helped Charlie to escape after the great Scots defeat at Culloden).

Back in Scotland, in 1779 Flora fell ill and was treated by Alexander Munro *secundus*, confusingly third of four generations of great doctors at the new Edinburgh University Medical Faculty. *Secundus* was third, because his father, also name of Alexander, was known as *primus*, because *his* father was named John. Lost? So am I. Anyway, in between dodging the bricks coming through his windows from an irate populace (who didn't like the way his students were raiding graveyards to snitch the corpses of their loved ones for dissection lessons), A.M. *primus* taught a Navy type, James Lind, who was the guy who came up with another term of opprobrium used to refer to the Brits (this time, by Americans).

In May 1747, on the good ship HMS *Salisbury*, Lind carried out probably the first proper controlled trial in the history of clinical nutrition. For fourteen days he kept six pairs of scurvy patients on the same diet, but gave each pair a different medicine: cider, elixir vitriol, vinegar, seawater, a "medicinal paste," or oranges with lemons. The citrus fruit did the trick. In 1753 Lind published *A Treatise of the Scurvy*, as a result of which, years later, the Royal Navy started issuing lime-juice rations to sailors. Who then never got scurvy, but had to put up with being called "Limeys."

Lind had been inspired to his researches by the shock news of a British Navy expedition that had gone horribly wrong. In 1740 Captain George Anson had sailed from England with six ships and over a thousand men. His mission: to head for the Pacific and clobber the Spanish wherever he found them. He did so, in spades, attacking Spanish ports and ships, laying waste right and left in the usual manner, and coming home four years later with so much treasure it took thirty wagons to haul it from the docks to the Tower of London for safekeeping. Every crew member walked off Anson's ship rich for life. There was a lot more booty than originally planned for each man to share because, of the original six ships and one thousand crew, only one ship with 145 men made it back. Scurvy had killed the rest.

Ironically, it was another medical emergency that had been the reason for Anson's voyage in the first place. Back in 1731, the British brig *Rebecca* was in the Caribbean, selling contraband goods to passing Spanish galleons with more money than sense, when a bunch of newly invented coastguards turned up from nearby Havana and (in the words of a British letter of protest to the Havana government) left the brig in such a state "that she should perish in her passage." Miraculously, however, the *Rebecca* made it back to England, and seven years later (nothing happens fast in history) the captain, Robert Jenkins, was asked to appear before a parliamentary committee to tell his tale. He did so, flourishing a box in which he had kept his ear, ever since the Cuban coastguard skipper, one Juan de Leon Fandino, a well-known crazy, had cut it off during the skirmish. Why you would keep your ear in a box is beyond me, but if Jenkins hadn't, it might never have become an historic appendage. The parliamentary row that followed the demonstration of Jenkins's grisly evidence (and then the public furor whipped up about it) led to a conflict between Britain and Spain now known as the "War of Jenkins' Ear."

The French statesman Mirabeau later cited this conflict as a good example of what happens when you let war be declared by a bunch of politicos. After this ringing indictment of democracy, Mirabeau died. France was so grateful for everything he'd done (sorry, no time for details) that they renamed the Paris church he was buried in the "Pantheon." Which is where Léon Foucault hung his giant pendulum in 1851. I've mentioned the pendulum before. It swung in inertial space, and inspired the American weatherman with whom I began this essay to say: "If a body is moving in any direction, there is a force, arising from the Earth's rotation, which always deflects it to the right in the northern hemisphere, and to the left in the southern."

I guess that's one forecast you can really rely on.

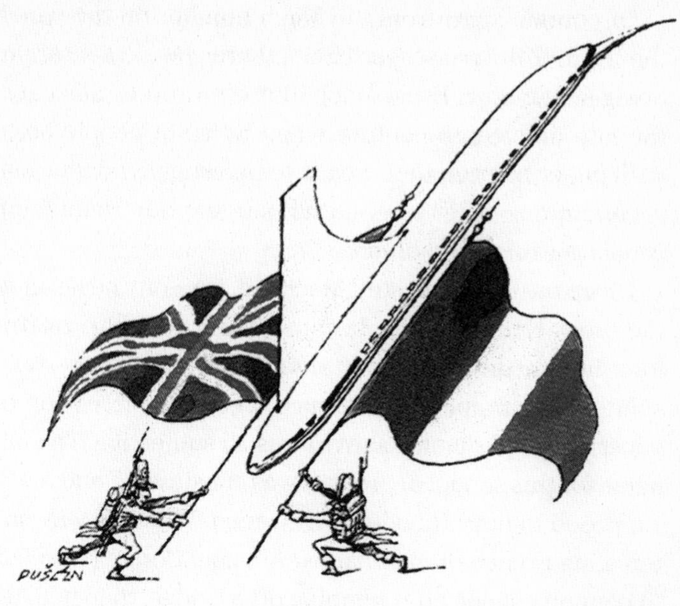

42 ENTENTE CORDIALE

IT WAS ONLY the other week I realized that the Anglo-French SST is called "Concorde" to celebrate Anglo-French amity. This hit me at sixty thousand feet over the Atlantic, as I watched everybody else on board pretending to be bored as we belted along at Mach 2.

Of course, apart from the Mach number on the panel at the front of the passenger cabin, there was no sensation of being supersonic. Ernst Mach himself foretold this back in the late nineteenth century when he spun people around with paper bags on their heads to investigate adaptation to acceleration. After the initial surge your semicircular canals get totally ho-hum.

Given the knockout stuff Mach did, it's a pity he never gets the press Einstein does, because Mach was into relativity long before the Great Man. For Mach there were no absolutes, just frames of reference, because perception (see paper bag experiments above) was all subjective. The fancy name for this school of thought was "Positivist," and Mach is supposed to have founded it in Vienna. Which is why an interesting French thinker named Auguste Comte (who failed to commit suicide after jumping off a bridge, then married a hooker, then started sociology) never gets the press Mach does. But let me short-circuit this before it gets out of hand: Mach got Positivism from Comte who got it from St. Simon who got it from Condillac who got it from Locke, who . . .

Anyway, Comte. The guy who first said the aim of science was prediction. And who split history into three eras: It's all gods; it's all mysterious forces; and it's all natural laws we'll formulate when we work out the math. Comte also came up with Social Physics: the way to predict behavior and make life easier for the zoners, planners, and politicians (give him a 3?). One of the other things Comte was positive about was the "general science of life" earlier developed by Marie-Francois-Xavier Bichat, who also boiled, fried, baked, dried, stewed, steamed, soaked, fricasseed (and in general applied the culinary act of reduction to) bits of animal and human, and identified twenty-one different types of tissue. And remarked that, since each tissue had different properties, each tissue would catch different diseases. You were looking for cause of death? *Cherchez la tissue.* Fortunately for devotees of whodunits and forensics in general, Bichat's bit of noodling became known as "pathological anatomy."

Of course this physiological bottom-line approach of Bichat's wasn't his at all. Such things almost never are. In the new, late-eighteenth-century nature-philosophy school, Bichat was only a follower. The leader being Friedrich von Schelling (well, I could argue that Schelling got it from Fichte who got it from Kant . . . but you'd throw this book away in a fit).

Nature-philosophy was all about a Grand Unified Theory before the Grand Unified Theory, postulating the existence of a fundamental substrate from which everything was made (like Bichat's tissue, which is why he went looking for it). In 1797 von Schelling also gave Romanticism a bit of oomph with the idea that Nature was all about opposites: north-south magnetic poles, acid and base, hot and cold, that stuff. And that the life processes were all about the constant struggle between these opposites. Out of this struggle would come "unification on a higher plane." This kind of blah-blah made late-model rationalistic Enlightenment thinkers tear their hair out.

Not so a Danish wigmaker's apprentice named Hans Christian Oersted, who went on to a degree in pharmaceutics (why not), and then got all excited about this Romantic nature-conflict business. Reckoning that the conflicting positive-negative "magnetical" nature of electricity could be made to conflict so much, then if you shoved a current down a very thin wire, it'd make a magnetic field. In 1820 Oersted did so. It did so. Five years later an English bootmaker, William Sturgeon, wrapped a live wire round a soft iron bar and the magnetic field that happened was strong enough to move things with. Like a telegraph receiver key, when bursts of current came down the line. Not bad for a bootmaker, right?

Sturgeon's electromagnetic obsession (what else would you call five hundred repeats of Franklin's kite-in-a-thunderstorm experiment?) got him a job as lecturer to the East India Company Royal Military Academy, thanks to a word from the school's math prof., Samuel H. Christie. Christie's father founded the art-auctioneer firm of Christie's, which may be why, as a young kid, Sam is said to have been great pals with Sir Joshua Reynolds. Sir J. was the artist's artist: eminent prez. of the Royal Academy, friend of the king, and portraitist *extraordinaire.* You name a mover and shaker, he painted his picture. *And* one young woman from Switzerland, about whom he is said to have had a bit of a thing. I've spoken of her before, so I won't waste your time now: Angelica Kauffmann, toast of London in 1770. She painted, too, though more houses than people, perhaps.

Kauffmann had spent a few formative months in Rome in the company of Johann Winckelmann, inventor of the history of art and esthetics and all that stuff well-bred young women do these days between graduation and children. Winckelmann was the guy who told Europe to go and look at Pompeii, Herculaneum, et al. not long after they'd been dug up, so people could get a feel for what it must have been like back in A.D. 0, and in this way understand Classical art. Others, like Giambattista Piranesi, just drew every ruin they came across. Piranesi showed his work to the Scottish architect Robert Adam, who then went home and turned lumpy British ancestral homes into bijou Greek and Roman villas, complete with the trimmings, some of which were done by Kauffmann.

Adam commissioned anything vaguely metallic from Matthew Boulton, who had all kinds of cutting and stamping machines in his Birmingham factory, where the rest of his time was spent running James Watt's life. Boulton had started out as a maker of shoe buckles. Not a bad idea before the era of laces. No fool, Boulton also got into the steam-powered (thank you, James) coin-stamping game just as they were thinking about issuing a complete new coinage. In 1797 he got two contracts: to make the new British copper money, and to set up a new Royal Mint at the Tower of London. Where they were then able to downsize the staff, because Boulton's new machines could strike two hundred coins a minute with only one minder in attendance.

The fine detail made possible by the switch to steam presses encouraged a new, more artistic approach to coin design. In 1814 the director of the Mint brought in a flamboyant Italian named Benedetto Pistrucci, who put St. George and the Dragon on the sovereign and crown coins for the first time. Pistrucci was able to be so deft with his designs because he used a new pantograph reduction machine that would reproduce the tiniest detail. On one occasion this detail included Pistrucci's full name, instead of the customary initials only. This lack of good manners and the fact that he was a foreigner meant Pistrucci never got the chief engraver's job he deserved.

However, a few years before he died, in 1850 Pistrucci delivered his designs for the medallion commemorating the Battle of Waterloo, the event that would sour Anglo-French relations enough so that over a hundred years later we'd still be busy with kiss-and-make-up gestures, like giving the SST a French name: Concord with an *e*.

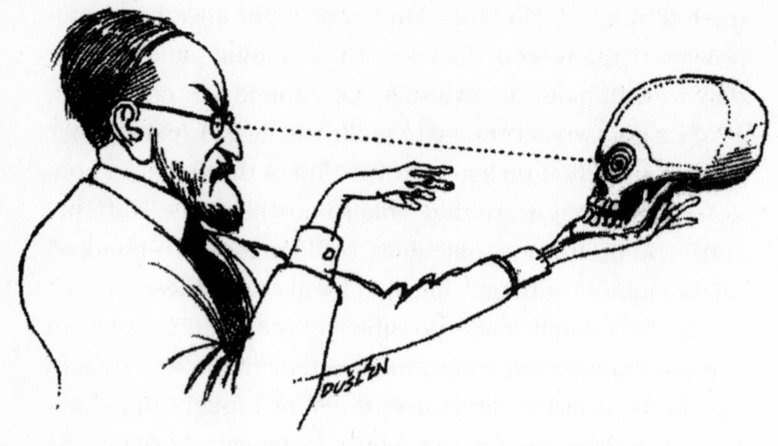

43
ZZZZZZZ

A RECENT BOUT of press hysteria about a stage hypnotist (whose subjects claimed he'd caused them long-term anguish) reminded me that's where it all started. With hysteria, I mean. When Josef Breuer, eminent Viennese medic, treated a lady named Bertha Pappenheim for what he thought was hysteria (symptoms: convergent squint, disturbances of vision, paralysis, contractures of arms and legs) with a radically new "gaze deep into my eyes" technique. The treatment worked so well that he and the colleague working with him kicked off the entire science of psychonanalysis. This made Breuer's colleague so famous I only have to say that his first name was Sigmund.

Freud shot off to Paris to bend the ear of anybody who'd listen, and seriously underwhelmed his prof., the hottest shot in French neurology, Jean-Martin Charcot (aka the "Napoleon of Neurosis," from the way he put one hand in his coat when demonstrating and was a general egomaniac and showman), because Charcot was too busy persuading the world that as far as the brain was concerned, mental was really physical. At this time the brain was considered the "most advanced" system of the body, controlling the nervous system and therefore responsible for all disease. Pretty much everybody since the Greeks had thought some kind of magic fluid ran down through the nerves from various parts of the brain to various parts of the body. Mind over matter, you might say.

Some time after 1810 Johann Kaspar Spurzheim and Franz-Joseph Gall (two Viennese doctors . . . what was it about Vienna?) had come up with a variation on this theme. Their idea was that the brain was composed of up to thirty-seven organs, each one of which controlled a specific personality characteristic. The more developed one of these control centers in the brain was, the bigger it was, and the more it made a bump on your skull that stuck out (a large bump behind the left ear meant you were a good lover, in case you want to check). In 1815 Spurzheim arrived in Edinburgh to lecture on grey-matter matters, and inspired two locals, George and Andrew Combe, to set up the Phrenological Society.

Phrenology was an instant smash hit (with, among others, Queen Victoria) because once you'd found the bump you were interested in you could maybe do exercises to make it even bigger. In the upwardly mobile self-improvement environment of the mid–nineteenth century, the ability to enlarge your bump of knowledge was irresistible. Phrenology even gave hope to social reformers, who wanted to reduce the size of criminal tendency bumps. For George Combe, things came to a head when he decided to splice the knot. So what could he do but examine the dome of his prospective bride? She passed the test and he married her because "her anterior lobe was large; her Benevolence, her Conscientiousness, Firmness, Self-esteem and Love of Approbation amply developed." She was also rich.

Even hard-headed (sorry, last cranial joke) businessmen fell for all this guff. Henry Maudeslay, a man so practical he invented the screw-cutting lathe without which the Industrial Revolution might not have happened, advised all young men to check out the skulls of their beloved (Maudeslay had also invented a machine that measured things to within one-ten-thousandth of an inch, so he was into quantification). One of Maudeslay's pupils was a young Napoleonic War draft-dodger named Richard Roberts, to whom I have alluded before for his invention of a machine that made rivet holes automatically on the Britannia Bridge and the SS *Great Eastern* steamship.

Early in his career Roberts had worked as a patternmaker for the king of iron, John Wilkinson, who ran one of the world's first blast foundries, in a place named Coalbrookdale (no prizes for guessing what you would see if you went there). It was Wilkinson who switched his furnace fuel to coal from charcoal, and started turning out pig-iron by the ton. Some time around 1770 he came up with yet *another* invention without which there might have been no Industrial Revolution: a machine that bored out cannon barrels from solid metal. Then, in spite of the state of Anglo-French relations at the time (down the toilet, heading for war), he smuggled the technology across the English Channel, whereupon the prerevolutionary French used it to make cannon, which they then shipped off to a bunch of people in the United States with a different kind of revolution in mind. Meanwhile James Watt jumped at the precision with which Wilkinson's machine could bore out cylinders accurately enough for his new engine to be steam-tight.

Wilkinson made enough money out of all this to get buried in an iron coffin (three times, till they got the right fit) and to provide finance and experimental equipment for his sister's husband, a failed preacher (he had a speech impediment) turned scientist named Joseph Priestley. Who hit the jackpot when he and Mrs. Priestley moved in next door to a brewery in Leeds. Well, you can't fail to notice carbon dioxide in such circumstances, can you, so Priestley put it in water, and invented soda. To be perfectly fair, he also discovered oxygen, wrote the definitive book on electricity, and became pals with every science star of the day, including Ben Franklin. This last relationship was not entirely to the taste of the English "king-and-country" mob, who then burned Priestley's lab down and forced him to leave for America in 1794.

One of Priestley's other equipment suppliers (and fellow member of the Lunar Society, a group of innovators and liberal thinkers who met at every full moon when the night roads were easier) was potter Josiah Wedgwood, whom you'll know if you're into formal crockery. Wedgwood made a million because he called a dinner set he designed "queen's ware." Social climbers by the thousands bought the stuff. And the Empress of Russia. Made Neoclassical all the rage, because it was based on the style of the vases and statuary and pediments and plinths (and anything else he could carry) pilfered from Pompeii and environs by Wedgwood's antiques-collecting friend, Sir William Hamilton, envoy plenipotentiary to the Court of Naples. Who sometimes turned up for meetings with the Lunatics.

Another Lunatic was Erasmus Darwin, well-known for boozing and having turned down George III's offer to take him on as royal physician. Wedgwood's eldest daughter married Darwin's son and ended up as the mother of Charles Darwin. About whom no more need be said, except that his cousin was Francis Galton, a man with a reported IQ of 200 and deeply into statistics. On one occasion Galton carried out a survey on the effectiveness of prayer, and on another, the body weight of three generations of British aristocrats (who says IQ is everything?). Galton is perhaps more (in)famous for having coined the term "eugenics." At one point he also became an enthusiastic member of the British Association for the Advancement of Science.

In 1853 one of the association's regular attendees, James Braid, wrote a paper with a riveting appendix titled "Table-Moving and Spirit-Rapping." As part of Braid's investigations of mental therapeutics, trances, and animal magnetism, he also discovered how to induce "a particular condition of mind and body" that he believed was good for the health.

Josef Breuer would one day agree. Braid named his new trick "hypnotism."

Have to stop now. My eyelids are feeling very heavy.

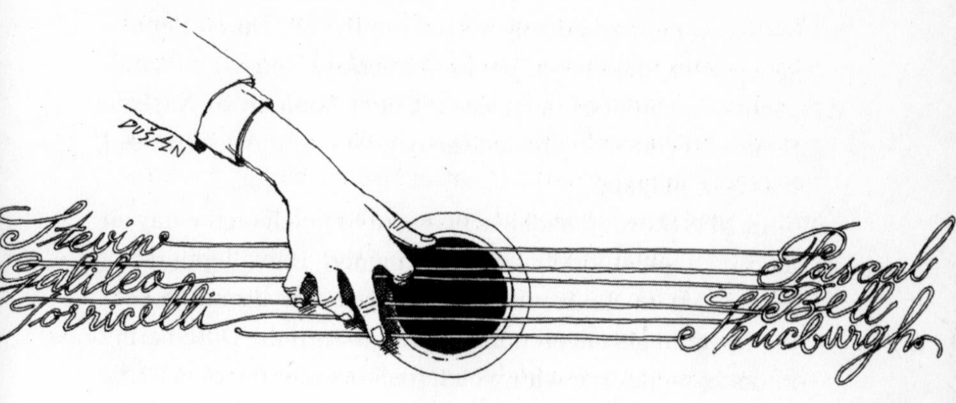

44

A FEW NOTES

ONE OF THE things I do to relax (it has the opposite effect on all within earshot) is to play the classical guitar badly, and the other day I warmed up by accompanying myself in a whistled rendition of "Yankee Doodle." I was then getting down to the serious matter of scales, to be followed by yet another failed attack on *Recuerdos de la Alhambra*, when it flashed upon my inward eye that musical scales had been mathematically sorted out by that Dutch Renaissance whiz, pioneer of decimal fractions, builder of sand-yachts and military adviser to Count Maurice of Nassau: Simon Stevin. Who divided the octave into the semitones I was now playing.

In 1608 Stevin must have been out to lunch on the day an unknown local optical person named Hans Lippershey fetched up at Maurice's palace with a new gizmo to help the count in his neverending efforts to turn the Dutch army into a hi-tech force with which to chuck out the occupying Spaniards (which he eventually did). Lippershey's gizmo was a tube with a lens at each end. To be used for "looking," as he put it. Maurice is reported to have muttered something about binoculars and sent him off with a flea in his ear. Next thing we hear, it's 1609, Galileo's got the kit and built one, and is about to change the entire history of everything by revealing that the Moon has mountains, and then proving the Earth isn't the center of the cosmos by showing moons orbiting some body other than the Earth: the planet Jupiter.

The solar-system concept was what Copernicus had been clobbered for by Rome nearly a hundred years earlier, so surprise surprise much the same was about to happen to big G. One of the other crimes Galileo was also to commit (which was almost as bad as heliocentricity) was to encourage researchers to do something about nothing. As in: the vacuum. Which was not supposed to exist, since any empty bits of the universe were reckoned by Rome to be filled by God's presence. The shenanigans all started in

1630, when Galileo was approached on the problem of why suction pumps wouldn't lift water more than about thirty feet. A serious matter when you were digging wells for water to power ducal fountains in Florence. Galileo kind of left the problem to one of his acolytes, Evangelista Torricelli (who lived in Galileo's house for the last years of his life and would end up succeeding him as mathematician and philosopher to the duke of Tuscany). Subsequent Torricellian thoughts led to a column of mercury in a tube being upended in a dish of mercury. During the upending, some of the mercury in the tube ran out into the dish, but some remained, in the form of a column of mercury reaching almost all the way up the tube. The gap above the almost? The impossible and as-you-know heretical vacuum. Over a number of days, Torricelli noticed that the level of the mercury in the tube crept up and down. Was this something to do with changing air pressure on the surface of the mercury in the dish which was supporting the column of mercury in the tube?

A copy of a secret letter about this risky idea, sent by Torricelli to a Roman propellor-head pal named Michelangelo Ricci, eventually found its way into the hands of the only person who was likely to be able to do something about things. This was Marin Mersenne, a scientific priest in Paris with the biggest address book in Europe. Mersenne was one of those guys who might not know the answer to something, but always knew a man who did. In this case, a guy who was now going to have to find a glass factory (one of the essential bits of vacuum experimental gear being long glass tubes). These were hi-tech materials, not readily available back then in your neighborhood mall. Unless you lived among the glassmakers of Rouen, which the guy did.

Then came the ticklish matter of mountains. Rouen is pancake-flat, and this chap wanted to take the mercury in the tube up a significant height to see if upness meant lower air pressure and a fall in the level of the column. Fortunately, the city of Clermont-Ferrand in central France had both mountains and his willing and able brother-in-law, name of Francois Peirier. On September 19, 1648, Peirier visited the mountain of Puy de Dome and went up and down. And so did the height of his mercury column, making like a barometer. And clinching the reputation of the guy who'd put him up to it all (4,888 feet up, to be exact): Mersenne's pal, and Francois' brother-in-law, Blaise Pascal.

Pascal was a mathematical genius who designed the first working calculator and was deeply into gambling and probability. Which may be why he was also in deep doodoo with Rome over his links with radical back-to-basics Catholic reformers known as Jansenists. These types, followers of Dutch priest Cornelius Jansen, attacked Jesuits for their probabilism ("What you're thinking about is probably not a sin if a church authority says not"). Jansenists, *au contraire*, were all for probabili*or*ism ("You never know—it's more probable that it's a sin than not, so don't do it"). Now back then, criticizing power-of-thumbscrew Jesuits was a good way to get yourself taken seriously dead. So by 1705, a papal bull required Jansenist priests and nuns to get out of the habit, or get out of the habit. Which is just what happened to Father Michel de L'Epée, who then went off to open a school in Paris, where he taught sign language to the deaf and dumb. Did so well, in a test one of his star pupils answered two hundred questions in three languages. By 1789 the school head was Roch-Ambroise Sicard, who finished the dictionary of signs L'Epée had started.

In 1815 an American, Thomas Gallaudet, turned up to learn the teaching technique. Two years later he had opened the Connecticut Asylum for the Deaf and Dumb in Hartford. In 1872 a Scottish immigrant did two months teaching at the asylum and then went off to Boston University to become professor of vocal physiology. It was there, while trying to develop a system to help deaf people to "feel" or "see" sounds, so as to imitate them, that the prof. took a close look at how the eardrum worked, and then found a way to make a vibrating membrane generate vibrating electrical current that would in turn vibrate another membrane. We call the resultant contraption Professor Alexander Graham Bell came up with the telephone. Since Bell was less than qualified to do any of the electrics involved, he wisely took advice from the eminent science guru and Smithsonian Secretary, Joseph Henry.

At one point early in his working life Henry had been tutor in the household of the Van Rensselaers, the Dutch patroon family that owned much of New York State in the seventeenth century. In 1642, in the city of Rensselaer, on the Hudson River, just across from the family home, Fort Crailo was built to protect the settlers.

Tradition has it that it was in this fort than an English medic, Richard Shuckburgh, composed that tune I was whistling at the start of this column.

Oh, well. Back to my scales.

45 SOUND IDEAS

I GOT A minor infection in one ear recently, and temporarily lost the ability accurately to locate sound sources. Made me really appreciate the way they used to plot the trajectories of incoming World War II rockets (aimed at me . . .

well, London . . . by you-know-who). British triple-A did the plotting with a number of separate microphones, each one picking up the sound at marginally different times, triangulating the sound of the engines. Consultant to these missile monitors was the youngest Nobel ever, physicist Lawrence Bragg, who'd earlier spent part of his World War I army service locating artillery by the same technique, known as sound ranging, which triangulated the sound of the bangs. Enemy gun positions were thus made as clear as crystal.

Which is what Bragg got his Nobel for. Crystals. In the summer of 1912 he and his dad (another Nobel) worked out how to tell the composition of crystals by bouncing X-rays against their atomic lattice. As the rays reflected (Bragg's term) off the line of atoms, they interacted with each other and created interference patterns that told you how the atoms were arranged and what the crystal was made of.

The basic technique had been worked out earlier that same year by a German named von Laue, who'd done it to prove that X-rays were very tiny electromagnetic waves (and would therefore mutually interfere if you bounced them off even tinier things like atoms). Von Laue made his interference patterns visible by exposing a photographic plate to them. The resultant image being known as Laue diagrams. All this minutiae had been inspired by a French expriest I've mentioned elsewhere: René-Just Haüy, who was talking to a colleague one day about a bit of calcite, when he dropped it. Noticing to his stupefaction that the fragments he'd turned the calcite into all looked remarkably similar, he took his little hammer and started smashing all the crystals he could find. Sure enough, what he would eventually describe as the "ultimate particles" of each type of crystal were all the same shape. In the case of calcite, rhombohedrons (as I'm sure you know). In 1801 this led Haüy to write the usual tome, establishing the science of crystallography, and stating that there were six basic crystal forms.

One particular German researcher took the news particularly hard. His name was Friedrich Mohs and he later argued (in 1822) that there were hardly *six* types. More likely *four.* Mohs's opinion on the subject had hardened while working hard to produce something that today is hard to avoid, any time a lady wants to check that her sparkler isn't paste. By which I mean that the well-known fact of a diamond being hard enough to scratch anything *less* hard is only well-known thanks to friend Mohs and his "Mohs hardness scale." In which Mohs ranked the hardness of ten materials, from talc:1, to diamond:10 (later additions to the scale included fingernails). Mohs's proximity to precious stones gave him an entree with the well-heeled, and he ended up counsellor to the Imperial German Exchequer, in charge of money matters.

In 1825 Mohs had a visit from a Brit who was keen to prep for a soon-to-be-available job of prof. of mineralogy. He got the job. Then, in 1841, the vice-chancellorship of Cambridge. Whereupon he dragged the university curriculum kicking and screaming into the nineteenth century. His name was William Whewell and I have a bit of a soft spot for him because he was a science popularizer and connectionist 150 years before I ever thought of it. Whewell was one of those Victorian polymaths: tidal expert, mathematician, writer of hexameter verses, German translator, Greek scholar, and inventor of the terms "ion," "anode," "cathode," "physicist," and "scientist." He also repudiated pointed arches and vaulting, in favor of flying buttresses, as the defining principle of Gothic architecture. And, if he hadn't been a clergyman, would have been a great boxer, they said. Whewell knew and organized the entire English scientific establishment and became the noodler's noodler.

As a schoolboy he'd taken lessons from "the blind philosopher," John Gough, up north in the English Lake District, where Whewell came from. Blind Gough was particularly good on math and botany—he felt plants with his tongue and lips—and also produced a mathematical theory of the speaking trumpet, studied ventriloquism, and, in an echo of my opening paragraph, investigated the "position of sonorous objects." Wordsworth and Coleridge thought Gough weathered his affliction remarkably well. An obsession with the weather was something he passed on to another pupil, John Dalton, who went on to make over two hundred thousand daily meteorological observations. In 1824 Dalton fell out of bed and died, after a last feeble entry: "Little rain this day."

Years of watching the behavior of water in the air naturally enough brought Dalton to an interest in the behavior of air (or any gas) in water. It was experiments to force various gases under pressure into water that then led Dalton to the startling thought (in 1803) that what he called "light, single" particles of gas absorbed into water less readily that heavier "complex" particles. The list of light and heavy particles he added to the end of a paper on the subject was the first of what we call the atomic weights table.

In 1792 Dalton was appointed professor at the Unitarian New College in Manchester, opened after the demise of the nearby Dissenters' Warrington Academy. Where Joseph Priestley had taught before being succeeded by Reinhold Forster. From 1772 to 1775 Reinhold and son George were the naturalists on HMS *Resolution*, when Captain Cook went looking for the hypothetical southern continent.

When they got back, the speed with which they brought out their book on the voyage beat the illustrious (and would-be author) Captain Cook to the punch. This put the Forsters in such bad odor with the harrumph naval establishment that George left for Germany. In 1790 he spent three months going down the Rhine with Alexander von Humboldt, presumably bending his ear with tales of naturalist derring-do from the great days of the Cook expedition. Whether Humboldt took notes or not, later on during his wanderings in South America, he did did much the same as had the Forsters in the Pacific.

Humboldt's writings blew away a footloose geographer and travel writer, Friedrich Ratzel, who then went off on a tour of the United States, studied the dwindling population of native Americans, and in 1901 came up with a theory about how population was related to space: the more of the latter, the more of the former. In 1921 Karl Haushofer, prof. of geopolitics at Munich, was teaching Ratzel's stuff to packed classes. Two years later he visited an ex-student who happened to be sharing a prison cell with a fellow who was writing up some great thoughts, and who jumped at Ratzel's "Lebensraum" theory of space because it accorded perfectly with his own ideas about the future expansion of Germany as a world power.

Haushofer's ex-student was named Rudolf Hess, and his fellow jailbird was the same guy who would later on be lofting over those V1s I mentioned at the beginning of this essay: A. Hitler.

Who also had his own hearing problems, as I recall.

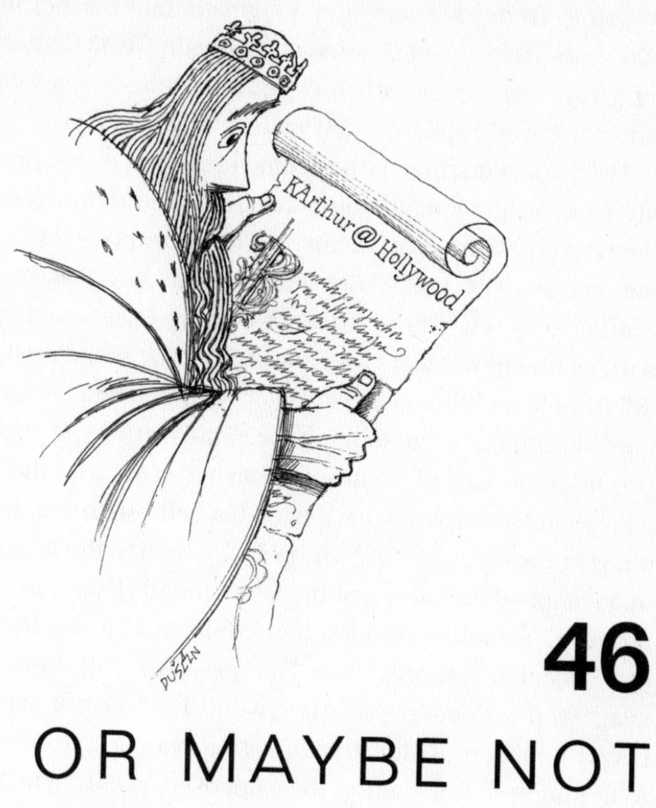

46

OR MAYBE NOT

WITH ALL THE academic research available these days about what it was *really* like, back in the Dark Ages when the European cultural lights went out (or maybe not), it's a pity Hollywood continues to churn out all that anachronistic garbage about King Arthur. You know: characters using terminology from 900 years later, knights in fancy armor from 700 years later, coats of arms and chivalry from 600 years later, turreted castles with drawbridges from 600 years later, riders using stirrups from 500 years later, and so on.

Mind you, clearing up these anachronisms would probably go over like a lead balloon at the box office. Which is how it went with one of history's greatest exposés of a similar nature. The box office in question was that of the Catholic Church, whose fifteenth-century boss was a pope with as much political clout as spiritual. Or so he thought. Till in 1440 a philological scribbler (aka humanist scholar) named Lorenzo Valla, looking for some dirt on the papacy (his boss the king of Naples was having a row with the Vatican about who ruled what), used his Latin smarts to point out that the language and terminology used in the hitherto-unquestioned document of the Donation of (Byzantine emperor) Constantine—which had conferred on the Roman pope secular authority over Europe—were (like the language and terminology of Hollywood King Arthur screenplays) bogus, *and* that the Donation was a fake, written four hundred years after the supposed event. Which of course put the kibosh on the pope's claim to temporal power. Everything curial hit the fan.

Valla only kept his head (literally) because he had a well-placed cardinal pal, the influential Nicholas of Kues, who had the papal ear and smoothed things over. Nicholas was a kind of Vatican ambassador-at-large, and the church's chief egghead. In mid–fifteenth century, independently of Copernicus, he opined that the Earth turned on its axis and wasn't the center of the universe. Also that there might be other inhabited planets. He advocated experimental methods (such as dropping things to measure their speed of fall and noting their air resistance) two hundred years before

Galileo. He talked about relativity five hundred years before Mach or Einstein.

Nicholas's big hero was a guy he'd met (when they were students at Padua University—the MIT of the time), name of Paolo Toscanelli, whom Nicholas described as the best mathematician alive. But Toscanelli was more than that, as he was to prove. To start with, after graduation he went home to Florence and told an architect friend all about the new Arab perspective geometry he'd been studying. The friend (Fillippo Brunelleschi) used the info to develop stuff like converging lines of sight and vanishing points, which excited an artist nicknamed Masaccio to kick off the whole of Renaissance art with his *Trinity* painting. Which was so realistic people thought they were looking at the scene through a hole in the wall.

By the time Toscanelli turned up at Nicholas of Kues's funeral in 1464, he was also deeply into cartography. He had read up on Marco Polo's trip and used Polo's data to work out the distance from Italy to Japan, which he then exaggerated to make his alternative route look better (that is, shorter by some ten thousand kilometers than it really was). At Kues's funeral Toscanelli talked the matter over with a Portuguese priest named Fernao Martins Roriz, who happened to head his country's permanent Commission on Exploration. Ten years later Toscanelli sent him a show-and-tell map for the Portuguese king. Who turned the idea down. So eventually Toscanelli offered the map to an Italian sailor keen to get to Japan, where they said the roofs were made of gold. And for whom Toscanelli's route west to Japan, across the Atlantic, with nothing in the way but water, was exciting enough to drum up the funds and go for it. On August 2, 1492, Columbus took his straight shot for Japan, headed for the biggest surprise in history.

That same day, others left Spain for very different reasons. To Spanish Jews, August 2 was shape-up-or-ship-out day, on which they were obliged either to turn Christian, leave Spain without their belongings, or be executed. Portugal was the nearest haven for one particular family, named Spinoza. Until 1580, when Spain (and the Inquisition) took Portugal over, and the nearest haven then became Amsterdam. So the Spinozas eventually fetched up there, and settled into the only truly tolerant country in Europe. In 1670 their philosopher descendant Baruch Spinoza strained even Holland's broadminded authorities with a publication calling for total freedom of thought and speech, denying miracles and the afterlife, and dumping religion in favor of numbers as the only way to explain the universe. By this time Spinoza's math had already attracted the attention of such Dutch science heavies as Christiaan Huygens, who introduced Spinoza to Henry Oldenburg, the English-speaking German who was secretary to London's Royal Society.

On behalf of which Oldenburg had set up a network of correspondents all over Europe and spent night and day writing and receiving letters about matters scientific. And, on occasion (when the writers inserted a bit of "for-your-eyes-only" espionage), *not* so scientific. This latter material Oldenburg passed on to the relevant authorities, as a result of which the Society was excused postal charges. One of Henry's other charges was Dora Dury (his ward), whom he married after his first wife died. Dora's father, John, was an Anglican clergyman who worked hard all over Europe to reconcile the various Protestant sects (he failed) and was, at one point, in Sweden trying to persuade Queen Christina to help (he failed). Christina had other things on her mind (like abdication), and besides, as an about-to-become-Catholic, she was the last person to ask.

Christina's intelligence was famous (she was known as the Minerva of the North), and while still monarch she would often invite known eggheads to come and stimulate her gray matter. One such guru was Hugo de Groot, Dutch legal eagle and the first (in 1609) to formulate a law of the sea, in which he said oceans belonged to nobody. This went over very badly with the English, who had recently clobbered a Dutch ship returning from Greenland with a cargo of twenty-two walruses on the grounds that the walruses were stolen property because Greenland waters were English. And codified as such, in a rebuttal of Hugo's case, by English megastar jurist and adviser to the king, John Selden. Who did himself a favor in 1618 with an essay dedicated to the new lord chancellor (top English lawyer), Francis Bacon.

Of whose science and philosophy output so much can be said I'll just note that he also wrote about the advancement of learning, in which he wanted everybody to share (he would have loved the Internet). One of Bacon's less famous observations was about how the continents on each side of the Atlantic seemed to fit together. It took until 1912 for a German meteorologist, Alfred Wegener, to come up with the explanation—continental drift.

Geologists pooh-poohed the idea for fifty years, sneering that Wegener was only a weatherman, and, not to put too fine a point on it, was seeing things. Interestingly, Wegener's other obsession was mirages. One of the most complex of which is known as "Morgan le Fay," after a famous witch of medieval legend.

Who had one other claim to fame. She was King Arthur's sister.

Or maybe not.

47

A MATTER OF DEGREE

I WAS ON the flight deck of a transatlantic triple-7 recently, watching the on-board navigational magic, when it occurred to me that it all went back to those two eighteenth-century French expeditions, sent to see if one degree of the meridian was longer up north than at the equator. Which it would be, if the Earth were an oblate sphere (flattened at the poles, as the English believed, and the French not).

The equatorial part of the mission involved sending a team to Peru, in the charge of intrepid Charles-Marie de la Condamine, who promptly discovered that the equatorial degree was indeed shorter. On his way home in 1743, Condamine rafted down the Amazon, scribbling busily as he floated along. One of the zillion things he described in passing was the *hevea* tree. Whose dried sap made a miracle substance that did something that, to eighteenth-century eyes, was weird and wonderful. It bounced.

By 1820 an English coach builder name of Thomas Hancock was buying all of this South American stuff he could get his hands on (not much) and making elastic waistbands and garters, soles and heels, false teeth, as well as all kinds of surgical trusses, belts, bandages, and so on. The market for rubber was soon insatiable, particularly when Hancock and partner Charles Macintosh spread it between sheets of cotton and invented the raincoat. So they wrote to the authorities saying, "Let's grow this stuff in our eastern colonies. Make a million, right?" The silence from Kew Royal Botanic Gardens (whose job such transplantations would be) was deafening.

Turned out the man in charge at Kew, William Hooker, was more concerned with a different tree, but from the same end of the world, and known as the *cinchona*, from whose bark quinine could be extracted. Point being that British imperial administrators and military types out in the steamier parts of the globe were dropping like flies from the effects of malaria. Harrumph. Quinine would put everybody back on their feet, so that the sun could continue never to set on the empire. Pip

pip. In 1852 the British government got a formal request to fund an expedition to pick up some *cinchona* seedlings, so Kew could nurture them till they were strong enough to be replanted in India. Alas, the little trees failed to grow well enough to solve the problem.

Meanwhile science wasn't coming to the rescue, in the person of William Perkin, a chemist, who fiddled around for several weeks in 1856 trying to make quinine, chemically. He finally came up with some black gunk that was definitely not quinine. So he chucked it down the sink, saw what it did to the water, and became a millionaire almost overnight, because he discovered that he had accidentally invented the world's first aniline artificial dye. As I'm sure you know, the raw material Perkin was using was the filthy-muck gaslight-manufacture by-product, coal tar, about which I have spoken before. Available by the ton, thanks to James Watt's sidekick, William Murdock, who kind of stole the idea of gaslight from Archibald Cochrane, eighth earl of Dundonald and amateur experimenter. Who produced coal-gas while making pitch to spread on navy ships' hulls to save them from boring teredo seaworms or something. The navy turned his pitch down (in both senses) and ruined him.

Ironically, Cochrane's son Thomas, the ninth earl, ended up a British navy admiral. This, after a checkered career that included head of the Chilean navy, head of the Brazilian navy, and head of the Greek navy. Thomas was also the inventor of the "Secret War Plan." Secret to this day. Cochrane claimed that his plan was capable of destroying any fleet or fortress in the world. In 1811 the Secret Plan was investigated for the British government by a Secret Committee, which turned it down on the wimpish grounds that it was "infallible, irresistible, but inhuman." So we'll never know.

One of the committee milquetoasts was William Congreve, whose own invention, the Congreve rocket, was to establish him in song and story. Well, song. As in "rocket's red glare," for it was hundreds of Congreve missiles that we Brits launched against Fort McHenry in 1814, exciting young Francis Scott Key to pen the present American national anthem. The music for which was, strange to tell, the work of an Englishman, John Stafford Smith, organist at the Chapel Royal. Back in England, in the 1770s, Smith was top of the charts with hits like "Flora Now Calleth Forth Each Flower." He was also more or less the first musicologist. Smith's boss at the chapel was composer Samuel Arnold, whose trick was to put together compilations of other people's stuff, add a bit of his own, and do very nicely. At various times, Arnold was also director of music at Covent Garden and Drury Lane.

Both theaters also employed the Spielberg of the period, David Garrick, who introduced the first high-tech special effects to the stage and brought realism to acting. An aristo patron of Garrick was Lady Dorothy Savile, a dab hand at caricatures, whose husband Lord Burlington was a mover and shaker in the art world. Burlington's live-in protégé (and the guy who taught Lady D. to draw) was William Kent. There are those who think Kent was a third-rate painter and a second-rate architect but a first-rate gardener. Well, maybe. His architectural magnum opus was Holkham Hall in Norfolk, said to be the first time an English architect had designed a house, interior decor, and furniture all in one. And people either love it or hate it. Pass.

The owner of the hall was Thomas Coke, earl of Leicester, who in 1822, at the age of sixty-nine, widower and father of three, married for a second time and then had six more. A man of breeding, you might say. Which he also made popular among farmers, in regard to sheep, pigs, and cows. All part of the agricultural revolution Coke helped to spearhead with other fancy practices such as crop rotation, turnips (they were used to feed livestock in winter), and clover (upped the yield because it nitrogenized the soil, although they didn't know it). Coke got many of his best ideas from people like Jethro Tull, whose 1731 book on husbandry was a rave seller in Britain and, twenty-odd years later, France.

Where one of Tull's most avid fans was the eminent thinker Voltaire, who went on to apply Tull's crop-improvement principles to his retirement plot at Ferney in Switzerland. This was well after the death of Voltaire's greatest love (out of many), Emilie du Chatelet, with whom he had spent a few happy years' bucolic intellectual idyll after they met in 1733 and recognized a common passion for Newton.

Emilie was learning algebra at the time, and for a while the three of them (she, Voltaire, and the algebra teacher) lived in a kind of *ménage a* $x + y + z$ in Emilie's château in Champagne. Then Z left on a trip, returning two years later in 1737, via Basel, where he picked up a young student who turned out to be such a lout that Emilie and Voltaire fell out with both of them. By this time Voltaire (like everybody else in France) was finding Z arrogant to a degree. Not surprising, given where Z had just been.

You remember I said one of those two French expeditions, headed by Condamine, went south to Peru, to work on geodetic matters? Well, Z (otherwise known as Pierre-Louis Moreau de Maupertuis) was the guy who went north, to measure the other meridian, up in Lapland.

48

ROOM WITH (HALF) A VIEW

I MAY HAVE mentioned earlier that I live in London, on the banks of River Thames within sight of the great Victorian railway bridge built by the great Victorian engineer Isambard Kingdom Brunel. He was half French, and that first sentence was half true, because one half of the bridge is obscured by the corner of the house next door.

As you'll know if you're a regular reader of my scribble, Brunel was the architect of the SS *Great Eastern*, the biggest steamship ever built at the time, which ended up successfully completing the final 1866 attempt by Cyrus Field to lay the transatlantic telegraph cable. Funnily enough, long before they ever thought of using *Great Eastern*, the ship was being built on one side of the Thames as they were putting together the cable on the other.

In preparation for the great event, Field had taken advice from Sam Morse, who'd already done something similar, though on a much smaller scale. Two years before he blew Congress away with his famous 1844 telegraph demo, he'd transmitted signals across New York Harbor with an insulated copper cable. This may have been why he'd also given a bit of cable (and presumably a few hints) to a neighbor who, that same year, wanted to blow up a ship just off the end of Manhattan by detonating a mine under it and persuading the U.S. Navy to buy the idea. The ship went down OK. And so did Sam Colt's fortunes when he wouldn't explain to the Navy how he'd done it. His revolver hadn't been doing too well either. Then along came the U.S.-Mexican War and suddenly Colt was back in the killing game. By 1855 he had the biggest private armory in the world.

His only rival in sudden death was the Remington Company, which ended up solving a major problem for Union troops in the American Civil War. Which was that while you were standing there ramming powder down your musket barrel, dropping in a ball, and then cocking and aiming, somebody shot you. Remington's breech-loading rifle changed all that and became the most successful military rifle in history. Sold more than a million to peace-loving nations all over Europe and the Middle East. After the Civil War ended, thanks to Remington for a while the pen became mightier than the sword when they turned over some of their now-idle machine tool lines to the production of a neat gizmo an inventor in Milwaukee had come up with. He'd done so after reading a description in the July 1867 edition of *Scientific American* of some Brit's attempt to do the same. Christopher Scholes's thing eventually became known as the Remington typewriter, and it freed women from kitchen drudgery so they could become involved in office drudgery.

The guy who had helped Scholes with his upper-case machine was an innovative legal type named Carlos Glidden. Noodling must have run in Glidden's family, because in 1874 a very distant relative of his named Joseph, living in De Kalb, Illinois, patented another device that was to become almost as popular with troops and farmers as was the Remington (rifle, not typewriter): barbed wire. Three years later Glidden sold his shares in the Barbed Fence Company to Washburn and Company Manufacturing of Worcester, Massachusetts. They were already producing Glidden's raw material, because by 1868 they were up and running with a new kind of wire-making mill, developed by a Brit named George Bedson. This would turn twenty tons of wrought iron into a zillion feet of quarter-inch wire in ten hours. A little earlier Bedson had also invented a continuous process for dipping wire into molten zinc and galvanizing it, so that, protected against wind and weather, it was suitable to be used by Ezra Cornell as he strung his telegraph wires all across America, thus making himself rich enough to found a university.

However, back to Washburn and Company. Around 1842 they turned down an offer from a young German engineer living in Pennsylvania who'd come up with a way to make wire rope by spinning the strands of wire on site. He'd had the idea when working on a curious system known as a portage railway. Before proper railroads superseded them, now and again canals would bump up against a mountain, and the only thing their builders (the young German was one) could do was to stick the canal barges on flatbeds and haul them on rails up and over the mountain to where the canal started again on the other side. Hauling was done with hemp hawsers. Which often broke. Hence the young German's wire ropes. Perhaps Washburn & Co. felt there just weren't enough canal-mountain interfaces to justify the idea. They must have kicked themselves when, in March 1855, the first train (carrying the Prince of Wales, and to a deluge of publicity) crossed the Niagara on a bridge suspended from the very wire ropes they'd rejected.

And imagine how they felt on the day in May 1883 when the whole of New York shut down for what was called The People's Day, and another bridge, once again hanging from John Roebling's wire ropes, was declared open, named one of the Wonders of the World, and finally United the States by linking Manhattan and Brooklyn.

Years before in Berlin, Roebling had apparently been persuaded to emigrate to America by his friend and Great Philosopher, G. W. F. Hegel. If you've ever had trouble with dialectical materialism, this is the guy to blame. Everything, Hegel said, contains contradictions within itself, and the tension between these contradictions is the driving force behind change, which happens when the contradictions are resolved. Geddit? In 1844 these musings changed the course of history when a twenty-four-year-old German journalist in Paris incorporated a version of them in *Economic and Philosophic Manuscripts.* One of those works you can't pick up, it talked about Hegel's tension in terms of class war, and resolution in terms of the inevitable triumph of the proletariat. Since the only safe place for this kind of madness in mid–nineteenth century was Britain, the author, Karl Marx, high-tailed it to London.

Where by 1884 the Executive of the Social Democratic Federation included Marx's daughter Eleanor. That year, when the SDF was infiltrated by anarchists, Eleanor decamped together with nine other members of the committee, including a wallpaper-maker and designer of rustic furniture named William Morris, who then founded his own, more democratic Socialist League. At Art Evenings held by the league in his London home, Morris read poems, George Bernard Shaw tinkled the ivories, and Chants for Socialists were sung by assembled members under the direction of one Gustav von Holst, an English-born trombonist.

Holst later dropped the "von" during World War I when he was put in charge of music for the British troops in Salonica and Constantinople. After the war he returned to fame and fortune with the first performance of the piece for which he is perhaps best known: the *Planets* suite.

I sometimes listen to it while looking out at my half-view of Brunel's bridge, partially blocked by the corner of that house I mentioned at the start. The one in which Gustav Holst lived.

49
VARIOUS, UNREQUITED

RECENTLY, IN THE beautiful Italian coastal town of Lerici, near La Spezia, I was staring out of the window of the Hotel Shelley at the bay where the eponymous poet drowned overboard in 1822, and thinking of his hard-done-by wife.

A year after the incident, pale and interesting Mary, passing through Paris on her way back to London, was enraptured to see a drama production of something she'd written after being inspired by Sir Humphry Davy's chemistry lectures, though maybe not the way Davy might have intended. The play was *Frankenstein* and the French reaction was "rave." Mary herself was having roughly the same effect on a local Romantic novelist, frequently unrequited lover, and Parisian culture-vulture, name of Prosper Mérimée (leaf through a few pages of his *A Chronicle of the Time of Charles IX* and your insomnia will be cured). I've spoken of Mérimée before, because you can't cross the French nineteenth century without bumping into him, since he made it his business to know everybody who was anybody.

One was a school pal: Jean-Jacques Ampère, Scandinavian-mythology freak, philologist, and son of André-Marie, the electrical whiz who gave his name to yet one more aspect of electricity I don't fully understand. Poor old Ampère *père* had a rotten life: guillotined father, early-death first wife, runaway second wife, and a mother-in-law who must have triggered the original jokes. So André-Marie buried himself in his work. Ended up shocking the intellectual live-wires of Europe with the science of electrodynamics. You get a feel for the man from the fact that while he was still a bookish child prodigy, some librarian in Lyons told him the mathematical text that he wanted to study that day was in Latin, so he went home and learned it.

Mind you, reading Daniel Bernoulli's math in *any* language must have been daunting enough. One of eight mathematicians produced by three generations of Bernoullis, each more incomprehensible than the next, Daniel turned out weighty stuff like the explanation for why your airplane's wings keep you up in the air (aka Bernoulli's Principle: with a curved airfoil the air over the [longer] top surface goes faster than the air under the [shorter] bottom surface, the over-the-top air pressure lowers, and you get sucked into the sky). Daniel also dipped into laterally discharged fluids, spheres rolling down inclined planes, the math of oscillating organ pipes, and (a fascinating glimpse into eighteenth-century high-tech) the optimal shape of sand-filled hourglasses. And his experimental physics classes in Basel laid 'em in the aisles.

One face in the student crowd was that of Joseph Frederick Wallet Desbarres, who then moved to England and in 1756 was commissioned as a lieutenant in the British Royal American Regiment, with the job of recruiting American colonials (who would be better than the Redcoats at fighting the French in the Canadian backwoods), and also of hiring anybody who could survey. Desbarres gave the job to Samuel Hollandt, ex-Dutch-army engineer and triangulator *extraordinaire.* Between them Desbarres and Hollandt (and a few helpers) began with local survey work in preparation for the British attack on Quebec and ended with the production of the great Atlantic Neptune charts. These gave the British Navy the triangulated lowdown on every creek and inlet along the entire, potentially revolutionary American Eastern Seaboard. Alas, by the time the last triangle had been dotted, so to speak, and the great work was ready for publication, it was already 1777. Too much, too late.

Back in 1759 Desbarres and Hollandt had taught most of what they knew about surveying (and its cousin skill, navigation) to a young naval nobody, who then went back home and became a somebody, ending up in command of HMS *Endeavour* on the 1768 Tahiti expedition to observe the transit of Venus. After which, during a total of three Pacific expeditions, he (Captain James Cook) charted much of New Zealand, discovered New Caledonia, the South Sandwich Islands, and South Georgia, and spent many months heaving up and down in the middle of oceanic nowhere. Back in Britain between voyages he became a big enough exploratory name to have his portrait painted by Sir Nathaniel Dance-Holland, whose main claim to fame (not much) was his brother George, a minor member of the new Neoclassical fraternity of architects. Once, that is, the two bros had taken the obligatory trip to Italy in the 1750s to ogle the recently uncovered and mind-boggling views of classical Pompeii, Herculaneum, et al.

Another local ogleworthy sight, with whom Nathaniel became unrequitedly infatuated (another one!), was a beautiful Swiss lady painter named Angelica Kauffman. At this time, ex-pat. *dolce vita* types in Rome (Ms. Kauffman included) sat at the feet of a gay Prussian art maven (well, he invented the study of art history), Johann Winckelmann, the oracle you went to hear (as did Nat and George), so as to find out what to see. And then what to say. Winckelmann wrote a massive analysis of ancient Greek art and architecture and was the first to suggest that you had to understand a period in order to understand its art. Historical relativism, I believe it's called now. Good stuff, except it laid the ground rules for the vacuous salon chatter that passes for cultural programming on TV and radio these days. Anyway, one of Winckelmann's many admirers was another Swiss, Henry Fuseli, a dauber who ended up in London and at one point edited a book by the Swiss divine Johann Lavater. This influential volume was titled *Physiognomy*, and featured the latest pseudoscientific stuff on how to read psychological traits from facial characteristics. Fuseli's own physiognomy must have been pretty good, because every woman he met fell for him. But the only meaningful, though (wait for it) *unrequited* relationship he seems to have aimed at (except for his marriage, of course), was with a woman named Mary Wollstonecraft.

She was probably one of the first true feminists, in 1792 writing a powerful emancipatory article: "A Vindication of the Rights of Women." In 1796 she had an affair with, became pregnant by, and married (in that order) one of the foremost liberal thinkers of the time: William Godwin. In 1793, when Godwin's *Political Justice* was published, it had made him an overnight celeb among free-thinkers of every stripe. In the tract Godwin advocated what sounds like an early form of communism and came down heavily on the side of nurture rather than nature, fingering education as the key element in shaping character. This went over very big with enlightened industrialists like Robert Owen, the first mill owner in Scotland to provide schooling for his mill workers' children (so he could mold them into good mill workers).

William Godwin was the radical's radical, even going so far as to say that *women* were capable of reason, a statement so outrageous, it did the trick with Mary. Alas (in the final moments of this tale of woe) it must be added that Godwin's passion, too, was fairly unrequited, since Mary Wollstonecraft Godwin died of a fever shortly after giving birth to their daughter. Leaving the child (named after her) to grow up without a mother's care and protection. Possibly as a result of which she was eventually to elope with a ne'er-do-well poet.

Who wrote good enough verse, but couldn't sail to save his life.

50 THE O ZONE

I WAS ON the beach a few months back (wearing sunscreen and hat of course) whiffing the tang of the sea air and thinking (as one does) about Christian Schönbein, who first discovered ozone (as the tang used to be called) in 1839 when he was playing around with electricity and water.

Seven years later he was to make a much bigger impression on the world with his other discovery. Arrived at by dipping cotton wool into a mixture of fuming nitric and sulphuric acids, then squeezing, washing, and drying it. The impressive result of which was that when you set light to it, the stuff went much more "bang" than gunpowder. Schönbein called the substance "gun-cotton," and in spite of the instant interest expressed by every military budget within earshot, one year later it was off the market for over a decade, because the first time an attempt was made to mass-manufacture it, the cotton blew up and totally obliterated the factory (and damaged large bits of the town of Faversham, England, a mile away).

In 1867 Shönbein's fatal fluff was destined to make a comeback as a result of the Great Disappearing Elephant Scare, when *The New York Times* predicted almost certain extinction if hunters went on bagging the beasts at the rate they were going. Billiards players faced a particularly grim future, since the best balls came from a perfect tusk, dead center. For which you need a plentiful supply of dead elephants. Harrumph. Which was why the firm of Phelan & Collander was offering ten thousand dollars for an ivory substitute, thus exciting the imagination of John Hyatt, a young printer in Albany, New York. In 1870 Hyatt mixed gun-cotton with alcohol and camphor, molded the result, and scooped the pool (well, you can't say "scooped the billiards"). The wonder material he came up with became false teeth, stiff collars and cuffs, vases, combs, fountain pens, dominoes, and about a thousand other things. I'm sure you've already guessed that it was also bound to find its way into cameras when in 1889 an ex-banker named Eastman patented photographic film made of "celluloid" (Hyatt's brother's name for fake ivory).

And here the plot thickens. Back in 1878 an English photographic weirdo who called himself Eadweard Muybridge (real name: Ed. Muggeridge) had taken a series of stills of a horse at the gallop to help Governor Leland Stanford of California win a bet about whether or not all the legs were off the ground at the same time (he said they weren't; he lost). These first action pix really turned on a Paris physiology prof., Etienne-Jules Marey, who was interested in the way anything that moved, moved. In 1887 he produced his *fusil chronophotographique*, which used a shutter to expose a roll of sensitized paper-based film to twelve shots a second. Two years later Marey showed this gizmo to Edison, who promptly bought some of Eastman's new celluloid. And in 1891 "invented" the cine-camera. Or (more probably) one of his unsung back-room noodlers did it. (Working to Edison's helpful laboratory motto: "There's a better way. Find it.")

It was another Edisonian egghead who also apparently came across an amazing high-vacuum pump developed a few years earlier by a German chemist, Hermann Sprengel. Word had also reached a Brit inventor, Joseph Swan, who was seeking the same kind of enlightenment as Edison: an incandescent bulb whose carbon filament wouldn't burn out if the vacuum inside the bulb were good enough. Which, thanks to good old Hermann, it now would be. And before we get into the "Edison or Swan" argument, I should point out that some other guy, in Cincinnati, had suggested incandescence way back in 1845. Anyway, in 1880, Swan installed the first of his lamps in the residence of Sir William Armstrong, a local politico, legal eagle, hydraulics engineer, field-gun designer, shipbuilder, and general manufacturing big cheese. The house was a spare-no-expense extravaganza named "Craigside," which Armstrong had designed himself and built in the middle of rave scenic surroundings as barons of industry were wont to in the days before zoning laws. All the place lacked (in common with everywhere else on Earth except Swan's or Edison's labs) was an electric light bulb.

Back in 1846 Armstrong had been elected to the Royal Society with the help of one Charles Wheatstone. And here we go again with the primacy thing. Together with a guy named Cooke, Wheatstone was the fellow who invented a telegraph in 1839, well before Morse (as, in some form or other, did about a dozen other people, too). Wheatstone's contraption worked by causing an incoming signal to deflect two magnetic needles to point at letters. Like all Victorians, Wheatstone did a lot more besides electromagnetics. He invented a coding machine, a way of indicating the sun's position from the polariziation of its rays, and the rheostat. But his real obsession was with acoustics. Not surprising. Wheatstone came from a family of musical instrument makers and in 1829 invented the concertina.

Without a doubt one of the greatest concertina exponents of all time (according to his family) was Lord Balfour, British prime minister from 1902, and later foreign secretary. When, in a letter to Lord Rothschild, he stated what became known as the Balfour Declaration, which gave official blessing to the plan for setting up what would eventually become the state of Israel. In 1921 Balfour became president of the Society for Psychical Research, joining such table-rapping pillars of the science establishment as physics professor Oliver Lodge.

Before these attempts at communication with the dead, Lodge also did much to improve the radio-telegraphic variety, when he developed (here we go *again:* as did Frenchman Edouard Branley) the "coherer." This device used metal filings to help detect radio waves because they stuck together when even a very weak electromagnetic signal passed through them. Marconi used the coherer to make possible the 1901 Newfoundland reception of his very first (very weak) transatlantic dots and dashes.

The question of how those signals had traveled round the bulge of the Earth was theoretically explained a year after the event by Charles Wheatstone's nephew, Oliver Heaviside. And, simultaneously, by American electrical engineer Arthur Kennelly (this is getting out of hand!!). Both men postulated some kind of stratospheric layer, off which radio signals might be bouncing. In 1912 one of Marconi's ex-assistants, a physicist named William Eccles, who had been involved in the preparation for the original transatlantic transmission, worked out a theory to show that a layer of ionized air would do that reflecting trick. In 1925 Appleton would prove Eccles was right with the discovery of the ionosphere, caused by the effect on the atmosphere of incoming solar X-ray and ultraviolet radiation.

However, only one year after Eccles had done his thing, in 1913 a Frenchman, Charles Fabry, was already fingering something else stratospheric and ionized. And also caused by the incoming ultraviolet radiation. It was a layer of gas that effectively shields life on Earth from the lethal effects of the same radiation. The shield is a little less effective today than in Fabry's time thanks to the hole in it. Which is why I was wearing all that protective gear on the beach.

Because Fabry found the same gas, up there, that Shönbein had discovered, down here.

SELECT BIBLIOGRAPHY

Allen, N. David Dale, *Robert Owen and the Story of New Lanark.* Edinburgh: Mowbray House Press, 1986.

Alvarez, M. F. *Charles V.* London: 1975.

Barchilon, Jacques, and Flinders, Peter. *Charles Perrault.* Boston: Twayne Publishers, 1981.

Batty, Peter. *The House of Krupp.* London: Secker & Warburg, 1966.

Beard, G. *The Work of Robert Adam.* London: Bartholemew, 1978.

Beatty, Charles. *Ferdinand de Lesseps, a Biographical Study.* London: Eyre and Spottiswoode, 1956.

Besterman, Theodore. *Voltaire.* Oxford: Basil Blackwell, 1976.

Blunt, Wilfrid. *The Ark in the Park.* London: Hamish Hamilton, 1976.

Bortoloan, Liana. *The Life and Times of Titian.* London: Hamlyn Publishing Group, 1968.

Bourde, André J. *The Influence of England on the French Agronomes, 1750–1789.* Cambridge: CUP, 1953.

Bowle, John. *John Evelyn and His World.* London: Routledge & Kegan Paul, 1981.

Bradley, Ian. *William Morris and His World.* London: Thames & Hudson, 1978.

Brockett, Oscar G. *History of the Theatre.* London: Allyn & Bacon, 1995.

Brooks, Jerome E. *The Mighty Leaf: Tobacco Through the Centuries.* London: Alvin Redman Ltd., 1953.

Brown, Pamela. *Henri Dunant.* Dublin: Wolfhound Press, 1991.

Brunschwig, H. *Romanticism and Enlightenment.* Chicago: University of Chicago Press, 1974.

Burke, James. *Connections.* New York & Boston: Little Brown, 1996.

Burke, Peter. *Montaigne.* Oxford: OUP, 1994.

Cameron, A. D. *The American Civil War.* Oliver and Boyd, 1985.

Cassirer, Ernst. *Kant's Life and Thought.* New Haven: Yale University Press, 1981.

Chancellor, John. *Audubon: A Biography.* London: Weidenfeld & Nicolson, 1978.

Chandler, D. G. *The Campaigns of Napoleon.* London: Weidenfeld & Nicholson, 1967.

Cole, Charles. *Colbert and a Century of French Mercantilism.* Hamden, Conn.: Shoe String Press, 1964.

Corbin, Diana. *A Life of Matthew Fontaine Maury.* London: 1888.

Davies, Ron. *John Wilkinson.* London: The Dulston Press, 1987.

De Sola Pool, Ithiel, ed. *The Social Impact of the Telephone.* Cambridge, Mass., & London: M.I.T. Press, 1977.

Douglas, Hugh. *Flora MacDonald: The Most Royal Rebel.* Stroud: Alan Sutton, 1993.

Driesch, Hans. *The History and Theory of Vitalism.* London: Macmillan & Co. Ltd., 1914.

Dunkel, H. B. *Herbart and Herbartianism: An Educational Ghost Story.* Chicago: University of Chicago Press, 1970.

Edwards, Owen Dudley. *The Quest for Sherlock Holmes.* London: Penguin Books, 1984.

Erickson, Carolly. *Bonnie Prince Charlie: A Biography.* London: Robson Books, 1989.

Fisher, Richard B. *Edward Jenner, 1749–1823.* London: André Deutsch, 1991.

Fitton, R. S. *The Arkwrights.* Manchester: Manchester University Press, 1994.

Fraser, Flora. *Beloved Emma, The Life of Lady Emma Hamilton.* London: Weidenfeld & Nicholson, 1986.

Gäbler, Ulrich. *Zwingli: His Life and Work.* Trans. Ruth C. L. Gritsch. Edinburgh: T. & T. Clark Ltd., 1986.

Gannon, Jack R. *Deaf Heritage in America.* Silver Spring, Md.: National Association of the Deaf, 1982.

Gernsheim, Helmut, and Gernsheim, Alison. *L. J. M. Daguerre. The History of the Diorama and the Daguerreotype.* London: Secker & Warburg, 1956.

Goldring, Douglas. *Regency Portrait Painter: The Life of Sir*

Thomas Lawrence, P.R.A. London: Macdonald, 1951.

Hall-Jones, Roger. *Jenny Lind.* Malvern: First Paige, 1992.

Halsband, Robert. *The Life of Lady Mary Wortley Montagu.* New York: OUP, 1960.

Hartcup, Adeline. *Angelica.* London: William Heinemann Ltd., 1954.

Hazlehurst, F. Hamilton. *Gardens of Illusion: The Genius of André Le Nostre.*

Herold, J. Christopher. *Bonaparte in Egypt.* London: Hamish Hamilton, 1962.

Hey, Colin G. *Rowland Hill, Victorian Genius and Benefactor.* London: Quiller Press, 1989.

Holmes T. W. *The Semaphore.* Ilfracombe, Devon: Arthur H. Stockwell Ltd., 1983.

Homer, W. I. *Seurat and the Science of Painting.* Cambridge, Mass.: MIT Press, 1964.

Honour, Hugh. *Neo-Classicism.* London: Penguin Books, 1991.

Hunter, James M. *Perspective on Ratzel's Political Geography.* Lanham: University Press of America, 1983.

Hutchison, Harold F. *Sir Christopher Wren: A Biography.* London: Victor Gollancz Ltd., 1976.

Hyman, Anthony. *Charles Babbage: Pioneer of the Computer.* Princeton, N.J.: Princeton University Press, 1992.

John, William D. *Pontypool and UK Japanned Wares.* Newport, Monmouthshire: The Ceramic Book Co., 1953.

Kardross, John. *The Origins and Early History of Opera.* Sydney: University of Sydney Press, 1957.

Kerby-Miller, Charles, ed. *Memoirs of the Extraordinary Life, Works and Discoveries of Martin Scriblerus.* New Haven: Yale University Press, 1950.

Knight, Frank. *Captain Anson and the Treasure of Spain.* London: Macmillan & Co. Ltd., 1959.

Knowles Middleton, W. E. *A History of the Thermometer.* Baltimore: Johns Hopkins University Press, 1966.

Lavine, Sigmund A. *Allan Pinkerton: America's First Private Eye.* London: Mayflower Paperback, 1970.

Lawson, Joan. *A History of Ballet and Its Makers.* London: Sir Isaac Pitman & Sons Ltd, 1964.

Leinwoll, S. *From Spark to Satellite: A History of Radio Communication.* New York: Scribner's, 1979.

Longford, Elizabeth. *Byron.* London: Hutchinson, 1976.

MacMullen, R. *Constantine.* London: Croom Helm, 1987.
Marriott, Ernest G. *Izaak Walton: A Short Study.* Nottingham: Nottingham Fly Fishers' Club, 1986.
Marshall, P. H. *William Godwin.* New Haven & London: Yale University Press, 1984.
McCormack, John. *One Million Mercenaries: Swiss Soldiers in the Armies of the World.* London: Lee Cooper, 1993.
McGlathery, J., ed. *The Brothers Grimm and Folktale.* Champaign: University of Illinois Press, 1988.
Mellor, Anne K. *Mary Shelley.* London: Routledge, 1988.
Miller, Edward. *Prince of Librarians: The Life and Times of Antonio Panizzi of the British Museum.* London: The British Library, 1988.
Moore, Doris Langley. *Ada, Countess of Lovelace.* London: John Murray, 1977.
Morton, S. G. *A Memoir of William MacClure.* Philadelphia: Academy of Natural Science, 1844.
Newton, H. W. *The Face of the Sun.* London: Pelican, 1958.
Norwich, John Julius. *The Normans in Sicily.* London: Penguin Books, 1992.
Ollard, Richard. *Pepys.* London: Sinclar Stevenson, 1991.
Phillips-Matz, Mary Jane. *Verdi.* Oxford: Oxford University Press, 1993.
Pierson, Peter. *Philip II of Spain.* London: Thames & Hudson, 1975.
Polnitz, G. von. *Anton Fugger.* Tubingen: 1958–86.
Roberts, Michael. *Gustavus Adolphus and the Rise of Sweden.* London: English Universities Press, Ltd., 1973.
Rolt, L. T. C. *The Aeronauts: A History of Ballooning.* Gloucestershire: Alan Sutton, 1985.
Rolt, L. T. C. *Thomas Telford.* Harmondsworth, Middlesex: Penguin Books Ltd., 1979.
Schuyler, Hamilton. *The Roeblings.* Princeton, N.J.: Princeton University Press, 1931.
Singer, Peter. *Marx.* Oxford: OUP, 1980.
Smith, Maxwell A. *Prosper Mérimée.* New York: Twayne Publishers, Inc., 1972.
Snyder, L. L. *The Roots of German Nationalism.* Bloomington: University of Indiana Press, 1978.
Stevenson, Edward Luther. *Willem Janzoon Blaeu.* 1914.
Stuyvenberg, J. H. van, ed. *Margarine: An Economic, Social*

and Scientific History. Liverpool: Liverpool University Press, 1969.

Taylor, A. J. P. *Bismarck, The Man and the Statesman.* London: Hamish Hamilton, 1955.

Taylor, Anne. *Laurence Oliphant, 1829–1888.* Oxford: OUP, 1982.

Trachtenberg, Marvin. *The Statue of Liberty.* London: Allen Lane, 1976.

Uerberhorst, Horst. *Friedrich Ludwig Jahn and His Time, 1778–1852.* Munich: Heinz Moos Verlag, 1982.

Van der Vat, Dan. *Stealth at Sea: The History of the Submarine.* London: Weidenfield & Nicolson, 1994.

Vaughan, Adrian. *Isambard Kingdom Brunel.* London: John Murray, 1991.

Viale, Mercedes. *Tapestries from the Renaissance to the 19th Century.* Milan: 1988.

Vogel, Dan. *Emma Lazarus.* Boston: Twayne Publishers, 1980.

Wason, K. *Delftware.* London: Thames and Hudson, 1980.

Watts, Michael R. *The Dissenters.* Oxford: Clarendon Press, 1978.

Wilton-Ely, J. *The Mind and Art of G. B. Piranesi.* London: Thames & Hudson, 1988.

Winegarten, Renée. *Madame de Stael.* Leamington Spa: Berg Publishers Ltd., 1985.

Wormald, Jenny. *Mary Queen of Scots: A Study in Failure.* George Philip: London, 1988.

Yovel, Y. *Spinoza and Other Heretics.* Princeton, N.J.: Princeton University Press, 1989.

About the Author

James Burke's books include the bestselling *Connections, The Pinball Effect, The Day the Universe Changed, The Knowledge Web*, *Tomorrow's World* (with Michael Latham and Raymond Baxter), and *The Axemaker's Gift* (with Robert Ornstein). He contributes a monthly column to *Scientific American* and serves as director, writer, and host of the television series *Connections 3*, which airs on The Learning Channel. He lives in London.